INTERNATIONAL CENTRE FOR MECHANICAL SCIENCES

COURSES AND LECTURES - No. 320

METHODOLOGY, IMPLEMENTATION AND APPLICATIONS OF DECISION SUPPORT SYSTEMS

EDITED BY

A. LEWANDOWSKI
WAYNE STATE UNIVERSITY

P. SERAFINI
UNIVERSITY OF UDINE

M. G. SPERANZA
UNIVERSITY OF BRESCIA

SPRINGER-VERLAG WIEN GMBH

Le spese di stampa di questo volume sono in parte coperte da
contributi del Consiglio Nazionale delle Ricerche.

This volume contains 37 illustrations.

ISBN 978-3-211-82297-5   ISBN 978-3-7091-2606-6 (eBook)
DOI 10.1007/978-3-7091-2606-6

PREFACE

In 1988 the idea arose within the group of people working in the Decision Science Program at the International Institute for Applied System Analysis (IIASA), Laxenburg, Austria, to held an international school to review the current state of the art in the subject of Decision Support Systems, with special emphasis to management applications. The presence of researchers from the University of Udine among these people and the aim of involving an Italian institution in this inititative, led naturally to the choice of the International Center for Mechanical Sciences (CISM), Udine, Italy, as a host institution.

After a preparatory phase the school was held in the week September 17-21, 1990 at CISM as a joint activity between IIASA and CISM. Also the Department of Mathematics and Computer Science of the University of Udine was partly involved in the organization.

The topics of the school were intended to be divided into general methodological issues and practical applications to various fields, like business, environment, transportation and production, with the goal, whenever possible, to show also the actual implementation during special software sessions.

Due to the broad scope of the subject a full week of lectures was planned by involving as lecturers internationally distinguished experts in the area. The school arose great interest and was successfull in all respects.

The present volume collects the proceedings of the school, although not all contributions could be prepared for publication. However, most of the lectures are reproduced here giving a faithful representation of the level and scope of the School. For ease of presentation the papers have been arranged in the book according to first author's alphabetical order.

The editors wish to express their deep gratitude first of all to CISM and IIASA. Without their organizational support the school could not even be planned. We are also deeply grateful to Unesco for its generous contribution which made possible the attendance to the school for many researchers from developing countries. Also we want to thank all the CISM's staff for its continuous and precious support which let all organization run smoothly. We also express our gratitude to all participants in the school whose friendly attitude established a nice and fruitful atmosphere.

Andrzej Lewandowski

Paolo Serafini

Maria Grazia Speranza

CONTENTS

INTELLIGENT DECISION SUPPORT SYSTEMS

H.W. Gottinger
University of Maastricht, Maastricht, The Netherlands
and
Fraunhofer Institute for Technological Forecasting, Euskirchen, Germany

H.-P. Weimann
Industrieanlagen-Betriebsgesellschaft, Ottobrunn, Germany

Abstract

This paper explores the basic ingredients of intelligent decision support systems in partial contrast to approaches followed by expert systems.

Rule based expert systems for decision support have been successful for well structured, well understood decision situations of a taxonomic classification type. But, in general, A.I. has growing influence in software engineering for ill-structured application areas by supporting an incremental development process with new programming techniques and architectures. As uncertainty is prevalent, information costly and payoff relevant, and the preferred solution depends on the specific beliefs and preferences of an individual or group decision maker the resolution methods of decision theory embodied in first-order predicate logic form a natural basis for computerized intelligent decision support. A unified characterization of knowledge and inference for logical, probabilistic, and decision-theoretic reasoning is developed for intelligent decision support over a wide spectrum of decision situations.

Key-words

Intelligent Decision Support, Expert Systems, Influence Diagrams, Decision Theory, Knowledge Engineering.

1. INTRODUCTION

In the past few years there has been substantial attention devoted to the use of artificial intelligence (AI) methods and architectures, most commonly rule based expert systems, as tools for decision support. An inherent focus of expert system development is the adequate modeling of human problem solving capabilities. In its sequel we observe the construction of several methods of representation like production rules, semantic networks, frames and scripts as well as inference mechanisms such as logic reasoning, non-monotonic reasoning and default reasoning, e.g., in facing problems like inconsistency and knowledge gaps. From the software engineering point of view, systems analysis can be done on a higher level of abstraction (closer to the domain expert) and involving the entire engineering cycle (Patrick 1986).

Especially rule based techniques have proven to be very attractive for a variety of problems particularly those which have fairly well structured (though possibly large) problem spaces, which can be solved through the use of heuristic methods or rules of thumb, and are currently solved by human experts. In these domains the reasoning and explanation capabilities offered by rule based expert systems are very effective. A rule-based approach tends to break down when applied to more difficult problems or problems that require a normative, prescriptive structure for decision and inference purposes, in particular, relating to the following situations:

- there is substantial <u>uncertainty</u> on various levels of decision-making;

- the preferred solution is sensitive to the <u>specific preferences</u> and desires of one or several decision makers;

- problems of <u>rationality</u> and behavioral coherence are intrinsic concerns of decision systems;

- problems of <u>resource-boundedness</u> for the user can be dealt with more adequately (Hansson and Mayer, 1988).

In established fields such as operations research and management science we have been developing methods for allocating resources under various conditions of time, uncertainty and rationality constraints. Central to these methods is the existence of an objective or utility function, as an indicator of the desirability of various outcomes. We will draw on this body of knowledge, especially elements related to the normative use of individual and group decision theory to approach difficult decision problems.

On the other hand, an evolutionary approach to system development is a major advantage of a production system or rule based program architecture and of expert system techniques in general. That is, once general decisions have been made regarding the basic control procedures and the organization of the rule base, the knowledge base can be incrementally improved by adding, modifying or deleting individual production rules. The advantage of rule-based program architecture combined with new programming paradigms such as object-oriented programming and logic programming facilitates advanced prototyping. In this light we develop methods for reasoning about the structure of probabilistic and decision theoretic models in a rule-based manner based on domain knowledge.

Summarizing, our attempt is to integrate conventional AI, logic-based approaches to problem solving with techniques for probabilistic analysis and decision making under uncertainty from operations research and management science to develop methods for improving the quality of decision making.

In this view, an intelligent decision support system (IDSS) is an interactive tool for decision making for well-structured (or well-structurable) decision and planning situations that uses expert system techniques as well as specific decision models to make it a model-based expert system (integration of information systems and decision models for decision support). The decision model imposes a normative profile on the IDSS serving for problem structuring and knowledge representation.

2. COMPUTER-AIDED DECISION MAKING

Advances in artificial intelligence, coupled with analytic techniques developed in the fields of systems analysis and operations research, can provide a means of significantly improving the quality of decision making by individuals and organizations. Traditional approaches to computer assisted decision making include decision support systems (DSS). The typical DSS provides means to sort, select, and transform information in the data base. Another recent development has been the use of artificial intelligent techniques, most commonly rule-based expert systems, as tools for decision support.

The efficacy of a rule based approach to decision support depends on the nature of the problem being solved. The classification-recommendation approach to decision making has significant limitations for particular types of domains. There are several incompatible taxonomies for the categorization of expert system problem areas. The most common scheme (Clancey, 1986) divides expert systems application areas into analysis problems (e.g. debugging, diagnosis and interpretation) and synthesis problems (e.g. configuration, planning and scheduling). Some problems cannot be classified that way because they comprise subtasks with many independent or semi-dependent sources of knowledge, interacting to find a common solution. The best example for such class of problems is speech recognition, but also in decision support we have problem areas such as planning/scheduling in military command and control. In those areas the "expert systems" approach supports the incremental improvement of the heuristic problem-solving process by adequate modeling of a rule- or frame-based representation embedded in an appropriate architecture, e.g. the blackboard architecture (Nii, 1986). Especially for the analysis problems, the heuristic classification based on an recommendation approach to decision making has significant limitations for particular domains.

Perhaps the biggest drawback of the "expert systems" approach is that most implementations do not have a general representation for the preferences or beliefs of the decision maker. This lack of a high level map to describe what the decision maker desires and believes has several ramifications. Another problem is that traditional AI systems are not well equipped to handle small differences in outcomes on a variety of attributes which may affect decision making. For example, most planning systems are based on developing a plan which can be proven to achieve a specific goal. The plan is either successful or unsuccessful (i.e. it can be proven constructively that there exists a successful plan or not), but most planning algorithms cannot evaluate trade-offs among factors such as the speed of achieving a goal versus cost and safety considerations. This contrasts with real decision situations, where alternative possible plans meet a variety of objectives to various degrees. In specific application areas combining analysis and synthesis problems (e.g. command and control) it might be advantageous to merge expert system techniques with decision procedures. In the planning and scheduling task for air traffic control or for the planning task of military operations for example we set up utility functions for the resource allocation process while we devise production rules and specific inference engines based on temporal reasoning (Allen, 1984) for scheduling activities. Finally, most significant decisions involve an element of uncertainty: that is, the decision maker lacks information about some aspects of his problem. Rule-based systems operate deterministically in their own reasoning, however, they can be engineered to be effective in a particular uncertain domain (the best example is the Mycin approach with uncertainty factors). In relatively well understood, static domains, one can design systems to look for the most likely cause of a fault before those that are less likely, and then recommend the repair strategy most likely to be effective. Therefore explicit treatment is not always necessary for a system which has to deal with uncertainty. However, in many cases uncertainty is encountered at a deeper level. Uncertainty arises because the situation is new or has not been previously considered.

Representations of uncertainty must be based on the information and beliefs of a particular decision maker, and cannot be delegated to an "expert". Finally, if there is an interaction between uncertainty, an individual's attitude toward risk, and the preferred course of action, then explicit consideration of uncertainty is needed.

3. INTEGRATION OF DECISION THEORY

We start out from recent efforts to design computer systems for decision support based on decision theory (Holtzman, 1985; Shachter, 1986). More general systems can be established by starting from group decision theory (team theory) for distributed decision making (Gottinger, 1989).

The basic result of the axioms of decision theory is the existence of a value function for scoring alternative sets of outcomes under certainty and a utility function for scoring uncertain outcome bundles. If the decision maker accepts the axioms (say, Savage's axioms; Savage, 1954; Gottinger, 1980) in the sense that he would like his decision making to be consistent with these axioms, then the decision maker should choose that course of action which maximizes expected utility. The importance of these axioms is that encoding decision procedures based on these axioms provide a basis for recommendations by an intelligent decision aid under uncertainty. They provide an explicit set of norms by which the system will behave. Other authors have argued why an individual should accept the decision axioms for decision making (Savage, 1954; Holtzman, 1987). The acceptance of these axioms is implicit in the philosophy and design of decision methods described here. The use of value and utility functions as criteria for decision making has several advantages. If the function is continuous with respect to outcomes, then it is able to handle small differences in outcomes in a consistent manner. This allows the computerized aid to handle an essentially infinite number of possible outcomes, not just those prespecified, foreseen, and categorized by the system's designers.

In addition, an approach to decision making based on decision theory has a mechanism, at least in principle, for handling completely new decision situations. The theory ensures the existence of a value and utility function. If the current expression of the preferences in the system does not incorporate the attributes of a new decision situation, the system can resort to the construction of a higher level or more general preference structure. By following these principles we are able to use the richness of modern decision theory and their axiomatic foundation (Fishburn, 1988). We could even encode ethical principles into decision theory (Harsanyi, 1976) and therefore enrich rational decision making in more than one dimension.

If the preference structure can be generated with sufficient generality, then the decision system can attempt to encode attributes of the new situation in terms of the general function, and use the new expression as a basis for decision making in the new situation. The task of developing robust preference models by incorporating deep and fundamental trade-offs is a difficult one. Development and elicitation of utility functions which reflect trade-offs regarding life and death issues as well as other dissimilar attributes is complex (Keeney and Raiffa, 1976). For the foreseeable future, assessment of utility functions for decision aids will necessarily be domain dependent. In fact, the applicability of decision aids such as those envisioned here will in all likelihood be limited by the ability to assess an appropriate representation of preferences. Domains in which there is a well developed empirical and theoretical basis for development of utility functions (e.g. financial and engineering decision making and some areas in medicine) are most promising.

Thus the decision axioms, along with the fundamentals of first order logic, provide a normative basis for reasoning about decisions. It is in this light that both logical and probabilistic inference will be utilized in an intelligent decision system.

So far we have distinguished between the problem solving capabilities of rule-based expert systems on the one hand, model-based decision recommendations using decision theory on the other. The choice of an appropriate technique or set of techniques for a given decision situation depends on many factors relevant to a particular decision. These include

- complexity of the situation,

- availability of alternatives,

- uncertainty with respect to the outcomes and relationships in the decision domain,

- strength of preferences with respect to alternative outcomes,

- magnitude of gain or loss possible in the decision, and

- requirements of procedural rationality and strength of heuristics available.

It seems reasonable to make all these characteristics an intrinsic part of a modelling process for an intelligent DSS design (Jarke and Radermacher, 1988). Modelling processes of this sort could possibly be more abstractly dealt with by "structural modeling" (Geoffrion, 1989) which establishes general principles for handling model-based resource allocation and decision processes.

4. DECISION MODEL BASED REPRESENTATION

For decision making, a model consists of the following elements (1) alternatives, (2) state descriptions (3) relationships, and (4) preferences. There can be no decision without alternatives, the set of distinct resource allocations from which the decision maker can choose. Each alternative must be clearly defined. State descriptions are essentially collections of concepts with which the decision is framed. It includes the decision alternatives and the outcomes which are related to the choices. The state description forms the means of characterizing the choice and outcome involved in the decision. The state description is also intertwined with expression of relationships. Relationships are simply the mappings of belief in some elements of the state description to others. The relations could be represented as logic relations, if-then rules, mathematical equations, or conditional probability distributions. The final component of a decision model is preferences. These are the decision maker's rankings in terms of desirability for various possible outcomes. They include not only his rankings in terms of the various outcomes which may occur in a decision situation, but also his attitude toward risky outcomes and preferences for outcomes which may occur at various times. They also embody information identifying those factors in a decision situation that are of concern, whether a factor indicates a desirable or undesirable outcome, and how to make tradeoffs among alternative collections of outcomes.

5. INFLUENCE DIAGRAMS

5.1 Basic Structure

As a computationally convenient way for a decision model based representation we deal with influence diagrams.We define the structure of influence diagrams (that in a superficial way resemble network flow representations; Ford and Fulkerson, 1962).

In other words, influence diagrams are network depictions of decision situations (Smith, 1988). Until recently, their primary use has been in the professional practice of decision analysis as a means of eliciting and communicating the structure of decision problems. Each node in the diagram represents a variable or decision alternative; links between nodes connote some type of "influence". Decision makers and experts in a given domain can view a graphical display of the diagram, and readily apprehend the overall structure and nature of dependencies depicted in the graph. Recently, there has been additional attention devoted to influence diagrams based on their uses in providing a complete mathematical description of a decision problem and as representations for computation. In addition to representing the general structure of a decision model, information characterizing the nature and content of particular links is attached to the diagram (Howard and Matheson, 1981). The diagram then presents a precise and complete specification of a decision maker's preferences, probability assessments, decision alternatives, and states of information. In addition the diagrammatic representations can be directly manipulated to generate decision theoretic recommendations and to perform probabilistic inference. The formalism of belief networks (Pearl, 1988) are identical graphical constructs which express probabilistic dependencies (no preferences or decisions). At this point, already, it is worthy to point out that inference and decision procedures using influence diagrams appear to be NP-hard (Cooper, 1987). Therefore, for some complex, multiply connected networks, it may be necessary to use approxiamtion algorithms. Approximation algorithms produce an inexact bounded solution, but guarantee that the exact solution is within those bounds.Following the notation of Shachter (1986) we define the syntax and semantics of influence diagrams.

Definition. An **influence diagram** is an acyclic directed graph G = (N,A) consisting of a set, N, of nodes and a set, A, of arcs.

The set of nodes, N, is partitioned into subsets V, C, and D. There is one value node in V, representing the objective of the decision maker. Nodes in C, the chance nodes, represent uncertain outcomes. Nodes in D, the decision nodes, represent the choices or alternatives facing the decision maker.

A simple diagram appears in Figure 1. By convention, the value node is drawn as a diamond, chance nodes are drawn as circles, and decision nodes are drawn as rectangles.

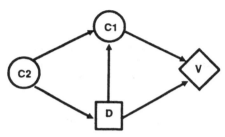

Figure 5.1 A Simple Influence Diagram.

V is the value node, the proposition which embodies the objective to be maximized in solving the decision problem. C1 and C2 represent uncertainties and D represents the decision. The semantics of arcs in the graph depend on the type of the destination node. Arcs into value or chance nodes denote probabilistic dependence. These arcs will be referred to as **probabilistic links**. Arcs terminating in decisions indicate the state of information at the time a decision is made.

Thus, C1 is an uncertainty which is probabilistically influenced (conditioned) by C2 and the decision. The ultimate outcome V, depends on the decision D and C2.

Definition . Each node's **label** is a restricted proposition, a proposition of the form (p t_1 t_2... t_n) where each t_i is either an object constant or alternative set.

We now define a set $\Omega(i)$ and a mapping π_i **for each node.**

Definition. **The set $\Omega(i)$** is the **outcome set** for the proposition represented by node i. It is a set of mutually exclusive and collectively exhaustive outcomes for the proposition.

Definition. The **predecessors** of a node i are the set of nodes j with arcs from j to i.

Definition. The **successors** of a node i are the set of nodes j such that there is an arc from i to j.

The mapping π_i depends on node type. The domain of each mapping is the cross product of the outcome sets of the predecessors of node i. Let the **cross product of predecessors** of i be **CP(i)** where

$$CP(i) = \{\Omega(i_1) \times \Omega(i_2) \dots \times \Omega(i_n) \mid \text{nodes } i_1, \dots, i_n \in \text{predecessors of node i}\}$$

The range of each mapping π_i depends on the type of node i.

5.2 Transformations

An influence diagram is said to be a **decision network** if 1) it has at least one node, and 2) if there is a directed path which contains all the decision nodes (Pearl, 1988; Howard and Matheson, 1981). The second condition implies that there is a time ordering to the decisions, consistent with the use of an influence diagram to represent the decision problem for an individual. Furthermore, arcs may be added to the diagram so that the choices made for any decision are known at the time any subsequent decision is made. These are "no-forgetting" arcs, in that they imply the decision maker 1) remembers all of his previous selections for decisions, and 2) has not forgotten anything that was known at the time of a previous decision.

The language of influence diagrams is a clear and computable representation for a wide range of complex and uncertain decision situations. The structure of dependencies (and lack thereof) is explicit in the linkages of the graph, as are the states of information available at each state in a sequence of decisions. The power of the representation lies, in large part, in the ability to manipulate the diagram to either 1) express an alternative expansion of a joint probability distribution underlying a particular model, or 2) to generate decision recommendations. The basic transformations of the diagram required to perform these operations are **node removal** and **arc reversal**.

These operations will be illustrated and defined with respect to a generic set of node labels: i and j are chance nodes, v is the value node. The labels p1, p2, and p3 will in general represent groups of predecessors of i, j, or v as indicated by the figures. In the interest of simplifying the descriptions of the operations, they will be treated as individual nodes. More detailed descriptions of the these operations appear in Shachter (1986), Smith (1988).

Removal of a stochastic chance node, **i**, which is a predecessor of a value node, **v**, is performed by taking conditional expectation.

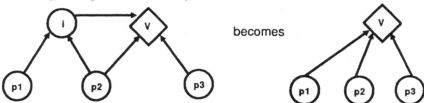

becomes

The new expected value function for v is calculated as follows:

$$\pi_{new,v}(\omega p1, \omega p2, \omega p3) = \sum_{\omega i \in \Omega(i)} \pi_{old,v}(\omega p1, \omega p2, \omega p3)\pi_i(\omega i | \omega p1, \omega p2)$$

The value nodes new predecessors are p1, p2, and p3.

Removal of a deterministic chance node, i, which is a predecessor to the value node, **v**, is performed by substitution. The picture of this process is the same as the previous case. The new expected value function for v is:

$$\pi_{new,v}(\omega p1, \omega p2, \omega p3) = \pi_{old,v}(\pi_i(\omega p1, \omega p2), \omega p2, \omega p3)$$

Removal of a stochastic chance node, i, which is a predecessor to another chance node, **j**, is also performed by taking conditional expectation.

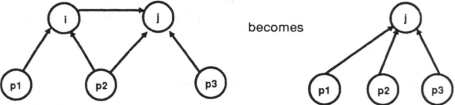

becomes

The new distribution for sucessor node j is calculated as:

$$\pi_{new,j}(\omega p1, \omega p2, \omega p3) = \sum_{\omega i \in \Omega(i)} \pi_{old,j}(\omega j | \omega i, \omega p2, \omega p3)\pi_i(\omega i | \omega p1, \omega p2)$$

The new predecessors of j are the predecessors of j other than i, that is, p1, p2 and p3.

Removal of a decision node, i, predecessor to the value node v is performed by maximizing expected utility. The decision node can only be removed when all of its predecessors are also predecessors of the value node; that is, the choice is based the expectations for the value, given what is known.

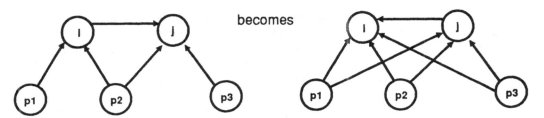

After removal the new expected value function for v is:

$\pi_{new,v}(\omega p2) = \underset{\omega i \in \Omega(i)}{MAX}\ \pi_{old,v}(\omega i, \omega p2)$

The new predecessors of v are the predecessors of i which are also predecessors of v, p2 as illustrated here. Note that there may be some informational predecessors of i, e.g. p1, which are not predecessors v before the removal. The values of these variables are irrelevant to the decision, since the expectation for the value is independent of their values. The optimal policy for the decision i is:

$\pi_i = \underset{\omega i \in \Omega(i)}{argmax}\ \pi_{old,v}(\omega i, \omega p2)$

This is the calculated π_i for decision nodes. We will refer to this calculated mapping as the decision function for i, $\pi_{d,i}(\omega p2)$. (See 6.3 Informational Influences).

Reversal of a probabilistic link between chance nodes is an application of Bayes' rule. Reversing a link from node i to node j can be performed as long as there is no other path from i to j (this is necessary to prevent the reversal from creating a cycle).

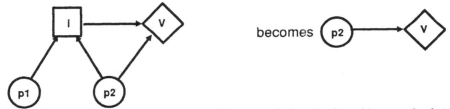

In reversing, the new conditional probability description for i and j are calculated as:

$\pi_{new,j}(\omega i|\omega p1, \omega p2, \omega p3) = \underset{\omega i \in \Omega(i)}{\Sigma}\ \pi_{old,j}(\omega j|\omega i, \omega p2, \omega p3)\pi_{old,i}(\omega i|\omega p1, \omega p2)$

$$\pi_{new,j}(\omega i|\omega j, \omega p1, \omega p2, \omega p3) = \frac{p_{old,j}(\omega j|\omega i, \omega p2, \omega p3)\pi_{old,i}(\omega i|\omega p1, \omega p2)}{\pi_{new,j}(\omega j|\omega p1, \omega p2, \omega p3)}$$

The operations of reversal and removal allow a well formed influence diagram to be transformed into another "equivalent" diagram. The original and the transformed diagrams are equivalent in two senses. First, the underlying joint probability distribution and state of information associated with each is identical, since the diagram expresses alternative ways of expanding a joint distribution into a set of conditional and prior distributions (Howard and Matheson, 1981). Secondly, the expectation for the value in the diagram and the sequence of recommended actions from decision node removal are invariant over these transformations (Shachter, 1986; Holtzman, 1987). In the next section, we focus on applying a sequence of these manipulations to obtain these recommendations.

5.3 Solution Procedures

On the basis of these manipulations, there exist algorithms to evaluate any well-formed influence diagram (Shachter, 1986). For purposes of probabilistic inference, we need two separate algorithms. In one version, which applies to well-formed diagrams, evaluation consists of reducing the diagram to a single value node with no predecessors, the value of which is the expected value of the decision problem assuming the optimal policy is followed. In the course of removing decisions, the optimal policy, i.e. the set of decision functions $\pi_{d,i}$ associated with each decision is generated. In the other algorithm, the objective is to determine the probability distribution for a variable, as opposed to its expected value. Both versions of the algorithm are described below.

Procedure **EXPECTED VALUE** (diagram)

    1. Verify that the diagram has no cycles.

    2. Add "no-forgetting" arcs between decision nodes as necessary.

    3. WHILE the **value node** has predecessors

        3.1 IF there exists a deterministic chance node predecessor whose only sucessor is the value node, THEN **Remove** the deterministic chance node into the value node

    ELSE

        3.2 IF there exists a stochastic chance node predecessor whose only sucessor is the value node, THEN **Remove** the stochastic chance node into the value node

    ELSE

        3.3 IF there exists a decision node predecessor and all the predecessors of the value node are predecessors of the decision node, THEN **Remove** the decision node into the value node

    ELSE

        3.4 BEGIN

            3.4.1 Find a stochastic predecessor X to the value node that has no decision sucessors.

3.4.2 For each sucessor S_x of X such that there is no directed path from X to S_x**Reverse** Arc from X to S_x

3.4.3 **Remove** stochastic predecessor X

4. END

At the conclusion of the EXPECTED VALUE procedure, the value node has no predecessors, and its single value is the expected value of the value node. Optimal decision functions are generated in the course of removing the decision nodes.

The algorithm to solve for a probability discription (or lottery) for a node is as follows:

Procedure **PROBABILITY-DISTRIBUTION** (diagram)

1. Verify that the diagram has no cycles.

2. IF the **value node** is deterministic, THEN convert to a probabilistic chance node with unit probability on deterministic values.

3. WHILE the **value node** has predecessors

   3.1 IF there exists a deterministic chance node predecessor whose only sucessor is the value node, THEN **Remove** the deterministic chance node into the valuenode

   ELSE

   3.2 IF there exists a stochastic chance node predecessor whose only sucessor is the value node, THEN **Remove** the stochastic chance node into the valuenode

ELSE

   3.3 IF there exists a decision node predecessor and all the predecessors of the value node are predecessors of the decision node, THEN **Remove** the decision node from the list of predecessor

ELSE

   3.4 BEGIN

      3.4.1 Find a stochastic predecessor X to the value node that has no decision sucessors.

      3.4.2 For each sucessor S_x of X such that there is no directed path from X to S_x **Reverse** Arc from X to S_x

      3.4.3 **Remove** stochastic predecessor X

4. END

The termination of this procedure is a probabilistic chance node with probabilities over the alternative possible outcomes of the original value node. Note that if decision predecessors are encountered in the algorithm, the distribution will be conditioned on the possible choice of the decision variables. The procedure does not remove decision nodes or generate decision functions.

6. DECISION LOGIC AND INFERENCE

The concepts developed above are now used to define a class of formulas (sentences) for decision domains. These formulas will be referred to as well-formed influences. A decision domain will be described in terms of a set of these well-formed influences.

A **well-formed influence** is defined as follows:

If A is a proposition and B is a conjunction of propositions B_i of the form $B_1 B_2 \ldots B_n$, then the expressions $A|_p B \equiv_\pi p(A|B)$, $A|_i B$ and $A\ B$ are well-formed influences.

A well-formed influence is an analog to a Horn clause in conventional logic programming. We have extended the Horn clause expression to incorporate statements about conditional probability distributions and to express information availability for decision making. A Horn clause in logic programming is a disjunction of propositions, in which all propositions but one are negated. A Horn clause has the following form:

$A \vee \neg B_1 \vee \neg B_2 \vee \ldots \vee \neg B_n$

or by De Morgan's Law $A \vee \neg(B_1 B_2 \ldots B_n)$ or in more familiar form as a rule $A\ (B_1 B_2 \ldots B_n)$ which is read IF B_1 and B_2 etc. are true then A is true. Also by definition, all variables in a Horn clause are universally instantiated; that is, the sentence is true for any term that is substituted for a variable appearing in the formula.

The choice of Horn clauses as a basis for influences is made for three basic reasons. First, there is a well understood, complete set of procedures for Horn clause logical inference. Inference is complete in that any Horn clause that is logically implied by another set of clauses is provable from that set using these procedures (Gallier, 1986). The inference procedures developed for Horn clauses will be the starting point for the probabilistic and decision theoretic techniques which are developed later in this section.

Secondly, first-order logic and Horn clause logic have proved to be an extremely expressive and useful language in a wide variety of situations. Expressing knowledge in Horn clauses is the basis for the logic programming language PROLOG, and forms the underpinnings for many derivative rule-based inference systems (Kowalski, 1979). First order logic is also the basis for several approaches to deductive databases and therefore supports the implementation process.

Finally, there is a natural parallel between the structure of a Horn clause and that of a conditional probability distribution, allowing a straightforward extension from logical rules to probabilistic rules. The Horn clause relates the truth of a proposition (A) to a conjunction of preconditions $(B_1 B_2 \ldots B_n)$ while a conditional probability distribution relates a probability distribution to a state of information - also expressed as conjunction of events. The representation of influences based on Horn clauses therefore rests on three pillars - computability, expressibility, and extensibility.

The proposition A (a single proposition) is referred to as the **consequent** of the influence and the conjunction B as the **antecedent** of the influence. We now describe in more detail the interpretation of each element of a well-formed influence (Breese and Tse, 1987).

6.1 Logic Influences

A logical influence is an implication formula of the form:

$A \leftarrow B$

This expression is a logical conditional, i.e., an IF-THEN rule. A logical influence with an empty antecedent is the assertion of a fact, i.e. a fact which is unconditionally true.

We can interpret logical statements in terms of zero-one probabilities. The correspondence between a logic statement of the form $A \leftarrow B$ and a conditional probability function π (A|B) can be developed.

6.2 Probabilistic Influences

A probabilistic influence, incorporating a probabilistic connector, |p, and a probability distribution is of the form:

$A \mid p \ B \equiv \pi_p(\omega_A \mid \omega_B)$

This sentence is the probabilistic analog to a deterministic logical influence. The left-hand side of the influence expresses the fact that the probability distribution over the alternative outcomes of A may be dependent on the outcome of B. The right hand side of the influence, $\pi_p(w_A \mid w_B)$, provides the numerical values of the distribution. It can be interpreted as providing the probability distribution over the outcomes of A for a given outcomes of B.

Just as logical influences express a means of asserting facts with certainty given other facts, a probabilistic influence expresses a measure of belief in a proposition given other facts. Suppose one desires to express the uncertainty of tomorrow's weather. Furthermore, we wish to condition the probability assessment for tomorrow's weather given the values of today's weather and a forecast of the weather of tomorrow. This can be expressed as follows:

(WEATHER x TOMORROW)|p (WEATHER y TODAY) (FORECAST z TODAY)

$\equiv \pi_p$ (ω(WEATHER x TOMORROW)|ω(WEATHER y TODAY) (FORECAST z TODAY)) where x, y, z ∈ {fair, cloudy, rainy} (the alternative set) and the w are members of the alternative outcomes for each proposition.

A probabilistic influence with an empty antecedent is the assertion of a prior (unconditional) probability distribution. For example:

$\forall$ y (WEATHER x y)|p ≡π p (ω(WEATHER x y)) =

ω	p(ω)
(WEATHER FAIR y)	0.3
(WEATHER CLOUDY y)	0.2
(WEATHER RAINY y)	0.5

Notice that in this example y is universally quantified - this is an assertion that for any value for y, the given probability distribution holds.

6.3 Informational Influences

An informal influence uses the informational connector |ᵢ to express information availability:

A |ᵢ B

This sentence denotes an informational influence and is only relevant in the context of decision making. The statement conveys two important pieces of information. The first is that A (a restricted proposition) is a proposition that is under the decision maker's control. The outcome of A from the set Ω(A) is not stochastic, but rather is selected by the decision maker. Thus as opposed to being the direct consequence of other outcomes (as in logic influences) or uncertain but conditionally dependent on other outcomes (as in probabilistic influences), its outcome is chosen from the set of alternative outcomes by the decision maker. Secondly, the propositions in B are known at the time the decision about A is made. Since the propositions may in general be probabilistic, this statement asserts that any uncertainties about their outcomes (within Ω(B)) will be resolved by the time a commitment on A is made.

Suppose the decision is the choice among alternatives for a rocket launch. The informational influence:

$\forall$ t((MISSION-CONTROL z t) |ᵢ (WEATHER y t))

where z ∈{LAUNCH,DELAY,CANCEL} and y ∈ {FAIR, CLOUDY, RAINY}

says that the launch decision for any time t is made knowing that day's weather.

An informational influence is purely descriptive. It in no way indicates what should be done in light of some other objectives. It differs from the other two types of influence in that they have some direct interpretation in terms of inference - knowing the antecedent of a logical or probabilistic influence tells something about the consequent, regardless of other information. Informational influences have inferential consequences only in the context of evaluating a situation for an optimal sequence of decisions. That is, only within the broader context of a decision model can one make a prescription regarding what choice **should** be made.

A function $\pi_{d,A}(\omega_B)$ is generated corresponding to an informational influence as a result of an optimization over the decision alternatives in terms of the decision model. In general, it will be a deterministic function mapping elements of $\Omega(B)$ - what is known - to the elements of $\Omega(A)$ - what can be done.

6.4 Decision Language

Recall the elements that are necessary to represent a decision domain - alternatives, state descriptions, and preferences. We will summarize by indicating how each element of a decision description can be expressed with respect to the constructs generated above.

First, recall that propositions form the basic unit of representation for a decision domain. There are three levels of knowledge regarding a proposition expressible in the language. First, it is possible to express a fact for a proposition, that is, a set of values for the variables (as in a fact substitution) in the proposition that are asserted to be true with certainty. Second, the values of the variables in a proposition may be restricted to some set. Thus, the outcomes for that proposition are restricted to a collectively exhaustive, mutually exclusive set, termed the alternative outcomes. Finally, a probability distribution can be used to associate each possible outcome with a probability. We have also shown how probability distributions and outcome sets are expressed for conjunctions of propositions.

Alternatives, the decision maker's options, are expressed in the set of outcomes for a proposition which is the consequent of an informational influence. The fact that a proposition has alternative outcomes and is the consequent of an informational influence defines it as a decision proposition. State descriptions consist of the set of facts and probabilities expressed within or deducible from a domain description. Relationships between states are expressed by the various types of influences available in the language; the logic, probabilistic, and informational influences expressed for the domain. Preferences are handled by identification of a particular proposition whose outcome incorporate the decision maker's objectives. A real valued variable in the proposition is identified as the objective - i.e. the value to be maximized or minimized. A logical influence is defined which ist capable of computing this value as a function of other propositions in the domain.

6.5 Example - A Decision Process

This section presents a simple example, using the decision language to describe a specific subproblem in a decision domain. Consider a security trader dealing in a single instrument, perhaps a particular Treasury security issue or foreign currency. The dealer's task is to trade continually in the instrument in order to make a profit. The trader's decisions are what quantity of the security to buy or sell at each instant of the trading day. The fundamental strategy is to "buy low, sell high", which is considerably easier to write down than to execute. The dealer's primary uncertainty is what the price of the security will be in the future. Changes in the price are dynamic and dependent on the price in previous periods as well as some other economic conditions or market factors. The trader wishes to maximize his expected profit at some terminal time (Cohen et al., 1982).

The following basic decision **alternatives** represent the trader's decision to buy, sell, or do nothing (hold) in each trading period. The set of propositions for this situation is shown below along with an interpretation for each. Alternative values for restricted variables are shown in brackets {}. These propositions constitute the means of expressing **state descriptions** for this domain:

(PROFIT profit time)	Trader's net profit. This is the cumulative total of all the trader's gains and losses in terms of profits since trading was initiated.
(POSITION value time)	Trader's net holding of the security. This is the cumulative total of all the trader's sales and purchases in terms of units of the security.
(TRADE {BUY SELL HOLD} time)	Trader's decision alternatives.
(PRICE {90 91 92} time)	Range of security prices. This is a restriction on the assumed range of prices that the instruments can adopt.
(FUTURES-EXPIRE time)	Futures contract expiration. Futures are contracts for the delivery of a given security at a future date. Standard security future contracts expire on a predetermined date (e.g. the 3rd Friday in March, June, September, etc.). This proposition is true if "time" occurs on a date when futures contracts mature.
(FUTURES-VOLUME {HEAVY MODERATE} time)	Indicator of acitivity level for futures markets. The level of activity in futures affects the levels of activity and prices in the "cash" market (i.e. for current delivery) that is considered in this example.
(GURU {BULLISH BEARISH} time)	Forecast by a market prognosticator or analyst. This represents the information of some outside expert. The "guru" is "bullish" if he believes prices ar likely to rise, and is "bearish" if prices are thought to fall.

We now describe the set of **relationships** which characterize this domain.

The trader's prift and position are simply accounting relations, expressed as determistic influences. We assume an initial position of zero units of the security, an initial profit of zero dollars, and a single trade quantity of 100 units. The facts (PROFIT 0.0 0)← and (POSITION 0.0 0)← indicate the trader starts with no holdings and no profit.

The net position of the trader is the difference between total sales and total purchases by the trader and depends on the trade made in the current period and net holdings in the previous period.

(POSITION new-position time)← (- time 1 last-time)∧

           (POSITION old-position last-time)∧

           (TRADE BUY time)∧

           (+ old-position 100 new-position)

(POSITION new-position time) ← (- time 1 last-time)∧

           (POSITION old-position last-time)∧

           (TRADE SELL time)∧

           (- old-position 100 new-position)

(POSITION new-position time) ← (- time 1 last-time)∧

           (POSITION old-position last-time)∧

           (TRADE HOLD time)

The profit level at any time is composed of the profit the trader has accumulated so far, plus an adjustment for the amount of the security the trader is holding (net position). If we identify maximizing profit as the objective of the trader, then his preferences among various outcomes (for PRICE, PROFIT, and POSITION) in terms of other propositions are expressed by the logic influence:

(PROFIT new-profit time)← (- time 1 last-time)∧

          (PROFIT old-profit last time)∧

          (PRICE old-price last time)∧

          (PRICE price time)∧

          (POSITION old-position last time)∧

          (- price old-price change-in-price)∧

          (* old-position change-in-price change-in-value)∧

          (+ old-profit change-in-value new-profit)

The PRICE proposition is the uncertain proposition in this example. The conditional probability distribution for PRICE is expressed by a series of probabilistic influences. A simple influence is:

(PRICE new-price time) $\mid_p$ (- time 1 last-time)∧

(PRICE old-price last-time)

$= \pi_p(\omega(\text{PRICE new-price time}) \mid \omega(\text{FUTURE-ACTIVITY level time}))$

This influence expresses a simple stochastic update of price given the previous period's price. If futures contracts for the security expire in a particular period, then the price is independent of the old price and is expressed by a different distribution:

(PRICE new-price time) $\mid_p$ (FUTURES-EXPIRE time)∧

(FUTURES-ACTIVITY level time)

$= \pi_p (\omega(\text{PRICE new-price time}) \mid \omega(\text{FUTURES-ACTIVITY level time}))$

Note that since (FUTURES-EXPIRE time) is not restricted, the conditional probability distribution π_p will not have separate entries for alternative outcomes of (FUTURES-EXPIRE time), therefore the distribution can be written solely in terms of the alternative outcomes for (FUTURES-ACTIVITY level time). The term (FUTURES-EXPIRE time) is a condition that must be true for the conditional probability distribution π_p () to be applicable. Note that the representation allows the expression of several conditional distributions simultaneously, and allows for the expression of conditions redarding which of the alternatives is appropriate by interweaving of deterministic information (e.g. FUTURES-EXPIRE) and uncertain outcomes (e.g. PRICE).

The trader also has information available from the market analyst (the "GURU") regarding his view of the market is being bullish or bearish. The guru therefore provides the trader with a indicator of overall market trends. The trader's opinion of this expert is expressed in the following influence:

(GURU assessment time p) $\mid_p$ (+ time 1 next-time)∧

(PRICE price time)∧

(PRICE next-price next-time)

$= \pi_p(\omega(\text{GURU assessment time}) \mid \omega \text{ (PRICE price time)}∧ \text{ (PRICE next-price next time)})$

This influence expresses the trader's probability distribution with respect to the guru's forecast for each possible set of prices for the current and subsequent period (i.e. the alternative outcomes of (PRICE price time) ∧ (PRICE next-price next-time)).

The final component indicates the decision in this domain and the information available. The influence states that at time 2 the trader knows the price and guru assessment.

(TRADE action2)| i (GURU assessment 2) ∧ (PRICE price 2)

For all periods, the trader knows the price when making a trade.

(TRADE action time)| i (PRICE price time)

These other facts and priors define the state of knowledge base at a given time. For example:

(PRICE price 0)| p ≡π_p(ω(PRICE price 0))

(FUTURES-ACTIVITY level time) | p ≡ π_p(ω(FUTURE-ACTIVITY level time))

(FUTURES-EXPIRE 2)←

The first two statements express prior probability distribution for prices at time zero, and futures activity for any period respectively. The last is a fact expressing that futures expire in period 2. The statements above, plus the values making up the probability distributions π_p() constitute the representation of this domain. The statements in this example are listed in the Appendix.

7. SUMMARY AND CONCLUSIONS

The techniques presented here are grounded on the premise that for effective intelligent decision support, the representation of a decision situation in a computer must reflect the alternatives, beliefs, and preferences of the user of the system. Therefore, the approach developed here focuses on 1) the development of representations and techniques which **construct** a probabilistic or decision-theoretic model for a particular user, query, and state of information, and 2) support the **exploration** of alternative representations and models for various phenomena by the user.

Central to the development of powerful computer-based decision aids is a means of expressing information and relationships important to describing a particular decision domain. We sketched a language based on first-order logic for the description of states, alternatives, beliefs and preferences associated with a decision domain and decision maker. The language includes constructs for explicitly enumerating the alternative possible outcomes for uncertain propositions, probability distributions over these outcomes, the choices facing the decision maker, and his preferences regarding alternative outcomes of differing likelihood. The concept of a logic rule has been generalized to provide for the expression of conditional probabilities (i.e. the probability of a proposition over its alternative outcomes given some conjunctions of propositions is true) and of information availability at the time of decision.

The current system relies on influence diagrams as a formalism for representing decision problems, and the algorithm developed by Shachter to solve for the optimal sequence of decisions. These techniques, while general, are inefficient for some problems, especially those which exhibit special structure that can be exploited in generating a solution. For example, recognition of separability of the value function in a Markov decision problem allows one to solve stochastic optimization problems using dynamic programming techniques. In addition, recent work on computational procedures for evaluating influence diagram (Chavez and Cooper, 1988) can also make the process more efficient in view of complexity bounds given by NP-hardness issues. In particular, KNET is a successful example of an IDSS that integrates decision networks and traditional expert systems. Other formal representations of decision problems, for example linear or non-linear programs, may be appropriate in some domains. Of particular interest are multi-person control or decision problems such as team theory or special organizational representations of command and control problems (Levis, 1984; Marschak and Radner, 1972) that are decision theoretic generalizations of the basic structure of decision theory (Savage, 1954).

The final basic area for future research is the incorporation of the ideas and concepts developed in this paper into artificial intelligence theories for autonomous rational agents. Currently, work in this area attempts to develop theories for belief, belief modification, goals, and action using classical logic and its extensions as a formalism (Georgeff, 1986; Rosenschein, 1985). Basing a theory of rationality on single person decision theory has several advantages:

- It provides an axiomatic basis for action.

- It insures the existence of a utility function providing a mapping from uncertain outcomes and decisions to preferences.

- It incorporates well grounded techniques for developing optimal strategies, handling uncertainty and risk preference, and calculation of the value of perfect and imperfect information.

- There exist well-tested methods and techniques application of decision-theoretic ideas to real world problems based on the professional practice of decision analysis and other system sciences (Holtzman and Breese, 1986; Howard and Matheson, 1984).

Such a theory should be flexible enough to allow for "bounded rationality" concerns (Simon, 1978; Gottinger, 1982). We should note that all decision making support by a computer aid reflects limited rationality to some degree. Even the most sophisticated DSS conceivable is limited in that it is based on a model, which by definition is an abstraction of reality and therefore contains inaccuracies due to cognitive limitations. The "perfect" or "complete" IDSS is an unattainable ideal.

REFERENCES

Allen, J. F., "Towards a General Theory of Action and Time", Artificial Intelligence, Vol. 33, No. 2, 1984.

Breese, J. and E. Tse, "Integrating Logical and Probabilistic Reasoning for Decision Making", AAAI, Third Workshop on Uncertainty in AI, Seattle, Wash., 1987, pp. 355-362.

Chavez, R. M. and G. F. Cooper, "KNET: Integrating Hypermedia and Bayesian Modeling," AAAI, Fourth Workshop on Uncertainty in AI, Minneapolis, Minn., 1988, pp. 49-54.

Clancey, W.J., "Heuristic Classification", Artificial Intelligence 27, 1985, pp. 289-310.

Cohen, J. B. and Zinbarg, E., Zeigel, A., Investment Analysis and Portfolio Management, Homewood, Illinois: Richard D. Irwin, 1982.

Cooper, G. F., "Probabilistic Inference Using Belief Networks is NP-hard", Report KSL-87-27, Medical Computer Science Group, Stanford Univ., Stanford, Ca. 1987.

Fishburn, P. C., Nonlinear Preference and Utility Theory, The Johns Hopkins Univ. Press: Baltimore, Md., 1988.

Ford, L. R. and D. R. Fulkerson, Network Flows, Princeton Univ. Press: Princton, N.J. 1962.

Geoffrion, A. M., "The Formal Aspects of Structural Modelling", Operations Research 37(1), 1989, pp. 30-51.

Georgeff, M., "The Representation of Events in Multiagent Domains," AAAI-86: Proceedings of the Fifth National Conference on Artificial Intelligence, 1986, Philadelphia, P.A., pp. 70-75.

Gottinger, H. W., Elements of Statistical Analysis, DeGruyter: Berlin, 1980.

Gottinger, H. W., "Computational Costs and Bounded Rationality," In Stegmüller, W. and W. Spohn, eds., Philosophy of Economics, Springer, 1982.

Gottinger, H. W., "Decision Making in Large Systems," International Conf. on Organizations and Information Systems, Bled, Yugoslavia, 1989.

Hannson, O. and A. Mayer, "The Optimality of Satisficing Solutions", AAAI, Fourth Workshop on Uncertainty in AI, Minneapolis, Minn., 1988, pp. 148-157.

Harsanyi, J., Essays on Ethics, Social Behavior and Scientific Explanation, Reidel: Dordrecht 1976.

Holtzman, S., Intelligent Decision Systems, Addison-Wesley, Reading, Mass., 1987.

Holtzman, S. and J. S. Breese, "Exact Reasoning about Uncertainty: On the Design of Expert Systems for Decision Support," In Kanal and Lemmer, eds., Uncertainty in Artificial Intelligence, Amsterdam: North Holland, 1986, pp. 339-346.

Howard, R. A. and J. E. Matheson, "Influence Diagrams," 1981, In Howard, R. A. and J. E. Matheson, eds., The Principles and Applications of Decision Analysis, SDG Publications, Strategic Decisions Group: Menlo Park, California, 1984.

Jarke, M. and F. J. Radermacher, "The AI Potential of Model Management and its Central Role in Decision Support," Decision Support Systems, 4(4), 1988.

Keeney, R. L. and H. Raiffa, Decisions with Multiple Objectives: Preferences and Value Tradeoffs, New York: John Wiley and Sons, 1976.

Levis, A. H." A Mathematical Theory of Command and Control Structures", Lab. for Information and Decision Systems, M.I.T., LIDS-FR-1393, Cambridge, Mass., Aug. 1984.

Marschak, J. and R. Radner, Economic Theory of Teams, Yale Univ. Press: New Haven, Conn., 1972.

Nii, H. P., "Blackboard Systems: The Blackboard Model of Problem Solving and the Evolution of Blachboard Architectures", AI Magazine 7(2), 1986, pp. 38-53.

Patrick, D., Artificial Intelligence. Applications in the Future of Software Engineering, Ellis Horwood: Chichester, 1986.

Pearl, J., Probabilistic Systems in Artificial Intelligence, Morgan Kaufmann, Los Altos, Ca., 1988.

Rosenschein, J. S. and M. R. Genesereth, "Deals among Rational Agents", Proc. Ninth Intern. Joint Conf. A. I., Los Angeles, 1985, pp.91-99.

Savage, L. J., The Foundations of Statistics, New York: Wiley Publications, 1954.

Simon, H.A., "How to decide what to do", Bell Jour. of Economics 8, 1978.

Smith, J. Q., Decision Analysis: A Bayesian Approach, Chapman and Hall, London 1988.

Shachter, R. D., "Evaluating Influence Diagrams," <u>Operations Research</u>, 34, 1986.

APPENDIX

Influences For Trading Example

(TRADE {BUY SHELL HOLD} time) ←   (A1)

(PRICE {90 91 92} time) ←   (A2)

(FUTURES-VOLUME {HEAVY MODERATE} time) ←   (A3)

(GURU {BULLISH BEARISH} time) ←   (A4)

(PROFIT 0.0 0) ←   (A5)

(POSITION 0.0 0) ←   (A6)

(POSITION new-position time) ←   (- time 1 last-time) ∧   (A7)
                                   (POSITION old-position last-time)∧
                                   (TRADE BUY time)∧
                                   (+ old-position 100 new-position)

(POSITION new-position time) ←   (- time 1 last-time) ∧   (A8)
                                   (POSITION old-position last-time)∧
                                   (TRADE SELL time)∧
                                   (- old-position 100 new-position)

(POSITION old-position time) ←   (- time 1 last-time)∧   (A9)
                                   (POSITION old-position last-time)∧
                                   (TRADE HOLD time)

(PROFIT new-profit time) ←   (- time 1 last-time)∧   (A10)
                                   (PROFIT old-profit last-time)∧
                                   (PRICE old-price last-time)∧
                                   (PRICE price time)∧
                                   (POSITION old-position last-time)∧
                                   (- price old-price change-in-price)∧
                                   (* old-position change-in-price change-in-value)∧
                                   (+ old-profit change-in-value new-profit)

PRICE new-price time) $\mid$ p (FUTURES-EXPIRE time)∧   (A11)

(FUTURES-ACTIVITY level time)

= πp (ω(PRICE new-price time) $^{\mid}$ ω(FUTURES-ACTIVITY level time))   (A12)

(PRICE new-price time) | p (FUTURES-EXPIRE time)∧

(FUTURES-ACTIVITY level time)

= πp (ω(PRICE new-price time) | ω(FUTURES-ACTIVITY level time))

(GURU assessment time p) | p (+ time 1 next-time)∧   (A13)

(PRICE price time)∧

(PRICE next-price next-time)

= πp(ω(GURU assessment time) | ω (PRICE price time)∧ (PRICE next-price next time))

(TRADE action2) | i (GURU assessment 2) ∧ (PRICE price 2)   (A14)

(TRADE action time) | i (PRICE price time)   (A15)

(PRICE price 0) | p ≡πp(ω(PRICE price 0))   (A16)

(FUTURES-ACTIVITY level time) | p ≡ πp(ω(FUTURE-ACTIVITY level time))   (A17)

(FUTURES-EXPIRE 2)←   (A18)

A DECISION SUPPORT SYSTEM FOR THE MANAGEMENT OF COMO LAKE:
STRUCTURE, IMPLEMENTATION AND PERFORMANCE

G. Guariso, S. Rinaldi
Politecnico di Milano, Milano, Italy

R. Soncini-Sessa
Università di Brescia, Brescia, Italy

Abstract

We review an application oriented study on the real time management of Lake Como, a natural multipurpose reservoir in Northern Italy. Emphasis is on the DSS that resulted from the study and has been used by the manager for several years to take his daily decision on the amount of water to be released from the lake. The structure of the DSS, based on the solutions of complex multiobjective deterministic and stochastic optimal control problems, is presented and the performance of the DSS during the first years of use is analyzed.

1. INTRODUCTION

This paper describes the basic findings and the practical implications of a five-year long research study on the management of a natural lake. The program was jointly supported by the Italian National Research Council (CNR) and by the International Institute for Applied System Analysis (IIASA).
The study focused on a specific reservoir, Lake Como in Northern Italy, for which a Decision Support System (DSS) was designed and implemented. Lake Como was selected because of the urgent need to prevent the increasing frequent flooding of the town of Como, which lies on the lake shore. In fact in 1946, when the regulation dam was put in operation at

the outlet of the lake, the level of the main square of the municipality of Como, which is the lowest part of the town, was much higher than it is now. Probably because of overpumping the water stored in the ground, the whole area has been sinking, so that flooding has become a very severe problem.

The paper is organized as follows. In the next section the main objectives and the legal and technological constraints of the management of Lake Como are analysed. In the following section, the DSS is briefly described, and then the operations it performs are reviewed in some detail in the rest of the paper. The most relevant part of the DSS suggests the amount of the daily release from the lake; this amount is computed according to three different optimization criteria (Section 4, 5 and 6). However, the DSS also performs inflow forecasting and lake simulation (Section 7), and data analysis and retrieval (Section 8).

2. THE MANAGEMENT OF LAKE COMO

Lake Como receives from a catchment of about 4500 Km^2, in the central part of the Alps, near Italy's border with Switzerland. The outflow rate of the lake can be varied from day to day by a manager who has the responsibility of operating the dam, built on the outlet of the lake (Adda River) at Olginate. The inflow rate averages 160 m^3/sec and has the typical annual pattern of alpine rivers with two peak flow periods, namely one in early summer due to snow melt, and another in autumn due to rainfall.

Water from Lake Como supplies a group of downstream users before reaching the Po river some 140 Km south of the lake. More precisely, 6 agricultural districts and 7 run-of-river hydro-electric power plants are located along the course of the Adda river and are served by a network of canals. The production functions of these users are not well-known and economic data on agriculture are scarce and quite unreliable. Thus, the only possibility is to use multiobjective optimization and to characterize the performance of the lake operation by means of simple physical indicators affecting the costs and benefits of all parties involved. The target of the downstream agricultural districts is the minimization of the total annual agricultural deficit A, i.e. the minimization of the water shortage (computed with respect to nominal requirements), while a good indicator for the power plants is the total annual electric power loss E (evaluated with respect to the installed capacity). Finally, the objective of the municipality of Como is the minimization of the number F of days of flood per year, which has been about 10 in the last 15 years.

The dam at Olginate is operated subject to the terms of a government licensing act, which specifies that the manager can decide the release each day provided that the lake level lies between two fixed limits x' and x" which identify the so-called "control range". This range multiplied by the surface of the lake gives the active storage, which is the volume of water the manager controls. The figure for Lake

Como is about 250 million cubic meters. When the level reaches the lower limit x' of the control range, the manager must release no more than the inflow into the lake, so that the level does not drop further to jeopardize navigation and cause sanitation problems. On the other hand, at the upper limit x" of the control range, the manager is obliged to open all dam gates to alleviate flooding of the lake shore.

Retaining water in the reservoir to irrigate agricultural land in the summer and to generate hydroelectric power in the winter runs the risk of disruptive flooding in Como and around the lake shore when the inflows suddenly increase. Thus, managing Lake Como is a process of making trade-offs between conflicting objectives: maximizing the water supply for irrigation and electric power generation and minimizing the risk of flooding.

3. THE DSS

To help the manager to take his daily decision, a set of programs has been implemented on a very cheap personal computer which has been utilized every day for a number of years, and has been recently expanded. The present computer is a standard PC with floppy and hard disks and a cassette tape for periodic data backup. All the data necessary for program execution are entered through a standard keyboard. The only exception is the rainfall data which are collected through an automatic telemetering network, that first comprised only two raingauges, but was later expanded to ten. The raingauges record rainfall every two hours and are connected to the computer through standard switched telephone lines. In normal conditions they are dialed up by the computer only once a day (their distance from the computer is about 100 Km) and they transmit all the raw data of the last 24 hours. Usually, this operation is performed every morning at 9 a.m. when the computer is turned on. Thus, in standard hydrological conditions only the total daily rainfall is computed from the raw data, and stored on the disk. However, during flood episodes rain gauges may be dialed up at any time and all raw data are stored in a separate file.

The software structure of the DSS is quite similar to that proposed by recent literature on environmental decision support system (see for instance: [1], [2], [3]) namely is composed by a data base and a model base. Furthermore, though not explicitly using expert system techniques, some of the models stored in the system and described below try to capture and formalize the experience gained by the managers of Lake Como during their 40 years long activity.

4. THE EFFICIENT OPERATION OF THE DAM

The first option provided by the DSS to assist the manager in the determination of the daily release r_t requires only two information items: the date t, and the lake level x_t at the beginning of the day. In other words, to take his decision the manager can use a function

$$r_t = r°(t, x_t),$$

called the efficient operating rule, which is programmed on his microcomputer.

This function was determined by off-line solving of a multiobjective optimization problem [4], [5] in which the objective was the minimization of the annual expected values of the three physical indicators A (water shortage in agriculture), E (electric power loss) and F (number of flooding days).

4.1. A stochastic optimal control problem

Formally, the problem was a multiobjective stochastic optimal control problem:

$$\min_{r(t,x_t)} | \; E[A] \quad E[E] \quad E[F] \; | \tag{1}$$

subject to

$$x_{t+1} = x_t + a_t - r(t, x_t) \tag{2}$$

$$0 \le r(t, x_t) \le L(x_t) \tag{3}$$

$$r(t, x_t) \le a_t \qquad \text{if } x_t = x' \tag{4}$$

$$r(t, x_t) = L(x_t) \qquad \text{if } x_t \ge x'' \tag{5}$$

and to a set of mass-balance equations representing the network of canals downstream of the lake and the actual rules of water distribution among the users.

In equation (1), $E[\cdot]$ represents the expectation operator: In the following the expected values E[A], E[E], and E[F] of the indicators will be denoted by $\underline{A}$, $\underline{E}$, and $\underline{F}$ for shortness of notation. Equation (2) is the continuity equation: The variation of the lake storage from one day to the next equals the difference between the inflow a_t, into the lake and the release $r(t,x_t)$ from the lake. In this equation all terms are expressed in centimeters (volumes divided by the surface of the lake) and a_t is a one-year ciclostationary stochastic process, namely a stochastic process in which mean and variance are periodic over the year. Equation (3) is the technological constraint which simply says that the release in one day cannot be greater than $L(x_t)$, the amount of water that can flow out from the lake when the gates of the regulation dam are kept permanently open; consequently the function $L(\cdot)$ is called the "open gates stage-discharge function". The last two equations represent the legal constraints imposed by the Ministry of Public Works: Equation (4) implies that the lake level cannot drop below the lower limit x' of the active storage, and equation (5) says that the dam must be completely open when the level of the lake is above the upper limit x" of the control range.

Many experiments of reservoir optimization have been reported in the literature in the last 20 years. The oldest contributions obviously refer to the simples problem: Finding the optimal operating rule of single-purpose reservoirs or of multi-purpose reservoirs in which, nevertheless, the different benefits and costs (associated with the degree to which the system objectives are met) are lumped together in a single objective function. More recently, a number of papers have considered the more complex problem of controlling single-purpose multi-reservoir systems or multipurpose reservoir systems. By and large, we can say that the propose solutions have only rarely been implemented by managers (see [6]). One reason is that in a number of cases the formulation of the problem and the solution algorithms are not transparent. A second and perhaps more important reason is that, in the case of existing reservoirs, the optimal solution is often very different from the one used in the past, so that the proposal of the analyst inevitably encounters the skepticism of the manager. In order to avoid these disadvantages we have solved the problem with a method that gives only suboptimal solutions; however, the manager is prepared to implement these solutions because they conserve those functional characteristics of the past decision making process that the manager believes to be important. The method proceeds in four steps.

(1) Conceptualization. Analyse the historical data on the operation of the reservoir, and with the help of the manager define the set P of those operating rules $r(\cdot)$ which are candidates for describing the decision making process followed in the past.

(2) Identification. Determine that particular operating rule $r^*(\cdot)$ within set P which approximates the historical releases best.

(3) Relaxation. Find some perturbations of the operating rule $r^*(\cdot)$ which might strongly affect the objectives, and check if these perturbations are acceptable to the manager. This analysis yields the set R of acceptable operating rules. This set contains the historical operating rule $r^*(\cdot)$ and all the operating rules $r(\cdot)$ obtained by relaxing $r^*(\cdot)$ in an acceptable way.

(4) Optimizations. Determine the solution $r(\cdot)$ of Problem (1)-(5) with the extra constraint that the operating rule $r(\cdot)$ belongs to the set R. Thus, if $r(\cdot) \neq r^*(\cdot)$ an improvement has been obtained by modifying the operating rule of the manager in an acceptable way.

While carrying out Steps 1 and 3 the analyst has considerable freedom in interpreting data, simplifying problems and suggesting solutions. Steps 2 and 4 however, are standard mathematical programming problems.

The first two steps of the method are described in detail in [7] and are summarized in [8]. The result of this analysis is that the identified historical operating rule

$$r_t = r^*(t, x_t)$$

is periodic over the year with respect to t, i.e.

$$r^*(t, x_t) = r^*(t+365, x_t).$$

Figure 1 displays this rule for any day t of the year. The figure shows that the control range [x', x"] is subdivided into three zones. In the lowest one (namely for x' < x_t ≤ $\tilde{x}_t$) the target is simply to satisfy agricultural requirement, i.e. the release r_t is equal to the agricultural water demand w_t or as close to it as possible; see the definition of L in equation (3). In the second zone ($\tilde{x}_t$ ≤ x_t ≤ x_t^*) demand for electricity production is also considered, while in the last zone (x_t^* ≤ x_t < x") flood protection becomes the dominant objective. The values x_t^* and r_t^* are given: therefore, each operating rule is uniquely identified by the parameters α_t, and β_t, t=1,...,365; see the angles α_t and β_t in Figure 1.

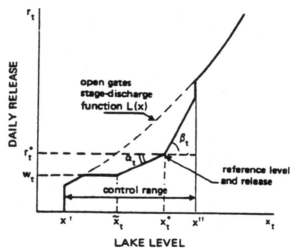

Fig. 1 - The operating rule of Lake Como.

The values x_t^*, r_t^*, α_t^* and β_t^*, t=1,...,365 of the historical operating rule have been estimated by minimizing the sum of squares of the differences between the historical values of the releases and the values which would have been generated by a systematic application of the operating rule shown in Figure 1.
The values α_t^* do not exhibit significant seasonal fluctuations, while the values β_t^*, which represent the sensitivity of the manager to floods, are strongly correlated with the pattern of inflows which peaks in early summer (snow melt) and autumn (rainfall). Furthermore, it turned out that $\beta_t^* > \alpha_t^* > 0$ for all t which explicitly quantifies the intuition that the operating rule is increasing and convex with respect to storage.
The next step was to define the set R of acceptable operating

rules (relaxation). According to the manager's recommendations, the values x_t^* and r_t^* or the reference level and release (see Figure 1) were left unchanged, while the values of the parameters α_t^* and β_t^* of the historical operating rule $r^*(\cdot)$ were considered to be modifiable, provided their seasonal patterns were preserved. Therefore, we considered as acceptable all the operating rules of the kind shown in Figure 1 with

$$\alpha_t = a\,\alpha_t^*, \qquad \beta_t = b\,\beta_t^*, \qquad t = 1,\dots,365,$$

where a and b are two unknown positive parameters to be determined through optimization. Thus, the set R of the acceptable operating rules

$$r_t = r(t, x_t, a, b)$$

contains the historical operating rule $r^*(\cdot)$, since $r^*(\cdot)$ corresponds to a=1 and b=1.

As is well known, the solution of Problem (1)-(5), with the extra constraint $r(t, x_t, a, b) \in R$ (which implies suboptimality but guarantees acceptability), is represented by a set of efficient (non-dominated) operating rules $r°(\cdot)$. Each one of them is identified by a particular pair $(a°, b°)$ of parameters, i.e.

$$r_t = r°(t, x_t) = r(t, x_t, a°, b°).$$

and has the property that any variation of such parameters deteriorates the expected value of at least one of the three indicators. In other words, if we denote by $\underline{A}(a,b)$, $\underline{E}(a,b)$, and $\underline{F}(a,b)$ the expected values of the indicators corresponding to an operating rule $r(t, x_t, a, b)$, we can say that the operating rule $r(t, x_t, a°, b°) \in R$ is efficient if there is no other operating rule $r(t, x_t, a, b) \in R$ such that

$$\underline{A}(a,b) \le \underline{A}(a°,b°),$$
$$\underline{E}(a,b) \le \underline{E}(a°,b°),$$
$$\underline{F}(a,b) \le \underline{F}(a°,b°),$$

with the strict inequality sign holding in at least one equation.

4.2 Numerical solution

From a computational point of view, the efficient solutions can be found very easily. In fact, for any given pair (a,b) of parameters a reliable estimate of the expected values $\underline{A}(a,b)$, $\underline{E}(a,b)$ and $\underline{F}(a,b)$ can be obtained by simulating the system for a sufficiently long period of time. In the case at hand, the simulations were carried out using the daily inflow data of the period from 1965-1979 (each simulation took about two seconds of CPU on a scalar mainframe). Thus, one can compute the values of $\underline{A}(a,b)$, $\underline{E}(a,b)$ and $\underline{F}(a,b)$ for a large number of grid points in the space (a,b) and then eliminate the solutions which are dominated.

In the case at hand, a grid of 60x60 points (a,b) around point

(1,1) (historical operating rule) was chosen and all the corresponding values $\underline{A}(a,b)$, $\underline{E}(a,b)$ and $\underline{F}(a,b)$ were obtained and stored in approximately two hours of CPU time. Then, a three dimensional graphic package was used on a color graphic terminal to visualize the surface

$$\underline{A} = \underline{A}(a,b), \quad \underline{E} = \underline{E}(a,b), \quad \underline{F} = \underline{F}(a,b)$$

in the space of the objectives, and the result was a shell shaped surface like the one in Figure 2. This figure clearly shows that the efficient solutions are those corresponding to the lower part of the surface. In fact, any point on the upper part of the shell, say point 1, is dominated by points on the lower part of the shell, say points 2, 3 and 4. The figure also clarifies that the efficient solutions are as infinitely many as the feasible solutions, which explains why it is appropriate to find them by an exhaustive search.

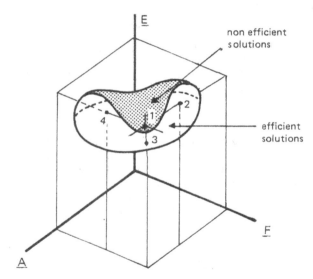

Fig. 2 - Efficient and non-efficient solutions
in the space of the objectives.

The point corresponding to a=b=1 lies on the upper part of the surface, thus indicating that the historical operating rule $r^x(\cdot)$ is not efficient and that the historical values of the objectives

$$\underline{A}^x = \underline{A}(1,1) = 201 \quad [10^6 m^3],$$
$$\underline{E}^x = \underline{E}(1,1) = 195 \quad [GWh],$$
$$\underline{F}^x = \underline{F}(1,1) = 10.2 \quad [d]$$

can be improved. The values of the objectives at points 2, 3, and 4, which correspond to three different efficient operating rules (a_2^o, b_2^o),

(a_3^o, b_3^o), and (a_4^o, b_4^o), are

$$
\begin{aligned}
&\underline{A}(a_2^o, b_2^o) = 79, &&\underline{E}(a_2^o, b_2^o) = \underline{E}^*, &&\underline{F}(a_2^o, b_2^o) = \underline{F}^*, \\
&\underline{A}(a_3^o, b_3^o) = \underline{A}^*, &&\underline{E}(a_3^o, b_3^o) = 190, &&\underline{F}(a_3^o, b_3^o) = \underline{F}^*, \\
&\underline{A}(a_4^o, b_4^o) = \underline{A}^*, &&\underline{E}(a_4^o, b_4^o) = \underline{E}^*, &&\underline{F}(a_4^o, b_4^o) = 5.1 .
\end{aligned}
$$

This proves, for examples, that by adopting the operating rules (a_2^o, b_2^o) or (a_4^o, b_4^o) one can substantially lower the agricultural water deficits or the floods without any negative consequences for the other objectives.

The exact figures of this multiobjective analysis are reported in Figure 3, which shows the lines in the space $(\underline{A}, \underline{F})$, corresponding to a constant value $\underline{E}$ of the electric power loss on the surface of efficient solutions. In the same figure the historical values $\underline{A}^*$, $\underline{E}^*$, and $\underline{F}^*$ of the objectives are shown together with the utopia point U, which represents the independent, and hence infeasible, absolute minima of $\underline{A}$ and $\underline{F}$.

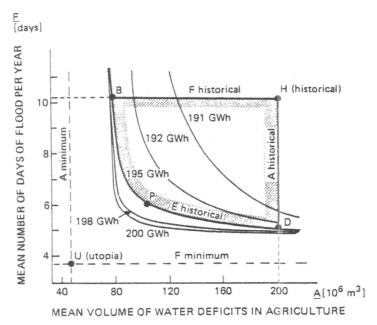

Fig. 3 - Efficient solutions in the space $(\underline{A}, \underline{F})$

We emphasize that the past management of the system can be improved by adopting any operating rule corresponding to points within the curvilinear triangle BHD. For example, selecting a point like P in Figure 3, which corresponds to an operating rule with $a^o = 1.95$ and $b^o = 2.10$, maintains the electric power deficit at its historical value and decreases the mean number of days of flood from 10.2 to 6.3 and the average agricultural deficit from 201 to about 100 million cubic meters (i.e. from 5.5% to 2.7% of the nominal annual requirements). The reduction of $\underline{F}$ is mainly due to the increase of the sensitivity to floods ($b°>1$), while the reduction of the agricultural deficit $\underline{A}$ is certainly due to the more conservative attitude toward droughts in the lowest part of the control range (the first zone becomes larger when $a° > 1$).

The efficiency of the operating rule with $a°=1.95$ and $b°=2.10$ was then verified by comparing all the historical droughts and floods with those that would have been obtained by systematically applying such a rule. A simple statistical analysis of all these floods and droughts (based on the Student distribution; see [5]) shows that one can be practically sure that the efficient operating rule ($a°=1.95$, $b°=2.10$) is better that the actual one. In fact, the 99% confidence intervals of the mean ratio between proposed and historical indicators of droughts and floods are the following:

        deficit volume: 0.32±0.17,
        deficit duration: 0.26±0.15,
        deficit peak: 0.69±0.27,
        flood duration: 0.30±0.16,
        flood peak: 0.87±0.03.

For this reason, immediately after this analysis was carried out, the operating rule was programmed on the manager's microcomputer.

5. THE ROLE OF ADDITIONAL INFORMATION

The DSS gives the manager a second option for determining the daily release r_t. The second option also requires information on the hydrometeorological conditions of the lake catchment. This information concerns

   y_t^1 = snow cover on day t,

   y_t^2 = depth of the aquifer on day t,

   y_t^3 = total rainfall during days (t-1) and (t-2).

Thus, to take his decision the manager may use a generalized operating rule

   $$r_t = r°(t, x_t, y_t^1, y_t^2, y_t^3)$$

which is programmed on his microcomputer.

The snow cover data are collected by the National Power Agency and transmitted to the manager only every 15 days, since the snow cover does not vary too quickly. Data on the aquifer depth are available whenever they are needed through normal telephone calls, while data on rainfall are automatically recorded every day through the telemetering network. Sometimes, in particular during winter, some of these data on the lake catchment are missing or very poor. For this reason, the generalized operating rule was a priori given the following structure

$$r^\circ(t, x_t, y_t^1, y_t^2, y_t^3) = r^\circ(t, x_t) + \delta_1^\circ(t, x_t, y_t^1) +$$

$$+ \delta_2^\circ(t, x_t, y_t^2) + \delta_3^\circ(t, x_t, y_t^3)$$

so that the contribution $\delta_i^\circ(t, x_t, y_t^i)$ could be disregarded when y_t^i is not available or is unreliable. Moreover, the function $r^\circ(t, x_t)$ was made identical to the efficient operating rule described in the preceding section. Therefore, each function $\delta_i^\circ(t, x_t, y_t^i)$ captures the value of the extra information y_t^i. In other words, by looking at the relative amplitudes of the terms $\delta_i^\circ(t, x_t, y_t^i)$ and by comparing them with $r^\circ(t, x_t)$ the manager will have the chance to learn more, in the years to come, about the real importance of these data.

The three functions $\delta_i^\circ(t, x_t, y_t^i)$ programmed on the microcomputer were determined by off-line solving of three independent multiobjective optimization problems (see [9]). Each one of these problems has the form (1)-(5) with $r(t, x_t)$ replaced by

$$r(t, x_t, y_t^i) = r^\circ(t, x_t) + \delta_i^\circ(t, x_t, y_t^i) \qquad (6)$$

where the function $\delta_i^\circ(t, x_t, y_t^i)$ belongs to a particular set D^i of admissible functions. In practice, this set of functions D^i was determined by very simple reasoning. More precisely, it was assumed that the generalized operating rule (6) could be formally obtained from the efficient operating rule $r^\circ(t, x_t)$ by making the reference level x_t^* and release r_t^* (see Figure 1) dependent upon y_t^i. For example, in the case of snow cover, whenever y_t^1 is greater than its average value $y_{t\ast}^1$ the reference level x_t^* is decreased, since less storage than usual is required in the lake if more resource (snow) is known to be available upstream. On the contrary, if less snow than normally is available in the mountains, the reference release r_t^* is lowered in order to store more water than is stored in standard conditions, so that the agricultural demand during the following dry season might be satisfied. Thus, the acceptable generalized operating rule $r_t = r^\circ(t, x_t) + \delta_1^\circ(t, x_t, y_t^1)$ is again shaped as the operating rule in Figure 1, with $\alpha_t = a^\circ \alpha_t^*$, $\beta_t = b^\circ \beta_t^*$ and the following variations in the reference schedule (x_t^*, r_t^*)

$$\Delta_1 x_1^* = -\varepsilon_1 (y_t^1 - \underline{y}_t^1) \quad \text{if } y_t^1 \geq \underline{y}_t^1 \qquad (7a)$$

$$\Delta_1 r_1^* = -\eta_1 (y_t^1 - \underline{y}_t^1) \quad \text{if } y_t^1 < \underline{y}_t^1 \qquad (7b)$$

where ε_1 and η_1 are two positive unknown parameters.

Problem (1) - (7) is again a multiobjective stochastic optimal control problem, but the operating rule has been parametrized, so that the independent variables to be optimized are only the two parameters ε_1 and η_1. In other words, the expected values $\underline{A}$, $\underline{E}$, and $\underline{F}$ of the three indicators depend upon ε_1 and η_1, i.e.

$$\underline{A}=\underline{A}(\varepsilon_1, \eta_1) \qquad \underline{E}=\underline{E}(\varepsilon_1, \eta_1) \qquad \underline{F}=\underline{F}(\varepsilon_1, \eta_1)$$

so that each efficient generalized operating rule is identified by a pair $(\varepsilon_1^\circ, \eta_1^\circ)$. Moreover, the efficient solutions are again represented by a two-dimensional surface in the three-dimensional space of the objectives (like in Figure 2). Thus, the efficient pair $(\varepsilon_1^\circ, \eta_1^\circ)$ can be obtained with the same procedure used to determine the efficient operating rule $r_t = r^\circ(t, x_t)$. The results are shown in Figure 4 in the plane $(\underline{A}, \underline{F})$ for hydropower deficit $\underline{E}$ smaller than or equal to 200 GWh. Point P in this figure is the same as in Figure 3. The three curves represent the best values of the objectives that can be obtained considering only one variable (snow cover, aquifer depth, rainfall) at a time. These curves show the improvements one can obtain with respect to point P (efficient operating rule described in the preceding section) by using only one piece of extra information. In each case the improvement is significant, particularly if compared with point U representing the utopic situation. The use of information on snow cover may, for instance, decrease agricultural deficits by 8.4% and floods by 17.7% of the maximum possible improvement (distance between points P and U). Similarly, the information on the underground aquifer may simultaneously decrease floods and agricultural deficits by about 13%. The information on rainfall, however, produces positive effects only on floods; but this decrease may reach 41% at the price of only 8.7% increase in agricultural deficit.

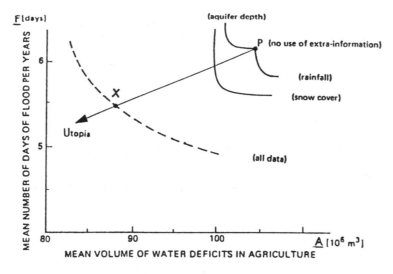

Fig. 4 - Efficient solutions when generalized operating rules are used.

If all the information (y_t^1, y_t^2, y_t^3) is available, a generalized operating rule

$$r^\circ(t, x_t, y_t^1, y_t^2, y_t^3) = r^\circ(t, x_t) + \delta_1^\circ(t, x_t, y_t^1) +$$

$$+ \delta_2^\circ(t, x_t, y_t^2) + \delta_3^\circ(t, x_t, y_t^3)$$

can be used where δ_1°, δ_2° and δ_3° might be any of the efficient operating rules corresponding to the curves in Figure 4. The best selection of all the possible combinations of the functions δ_1°, δ_2°, and δ_3° yields the dashed curve in the figure. This curve is only slightly different from the one obtained by a simple vectorial sum of the three solid curves. This is because the three items of information y_t^1, y_t^2 and y_t^3 are largely independent from each other.

So it can be concluded that the use of extra information on the catchment considerably improves the management of the lake. For example, point X in Figure 4 represents a 20% reduction of agricultural deficits and a 55% reduction of floods with respect to the maximum possible improvements. Therefore, immediately after these results were obtained, the generalized operating rule corresponding to point X in Figure 4 was programmed on the microcomputer and since then it has been used by the manager.

6. A RISK AVERSE OPERATING RULE

The third and last option the manager has for determining the release r_t on day t is quite different from the two preceding ones. In this case, the optimality criterion has nothing to do with the minimization of the expected losses. The criterion is risk-aversion, i.e. the minimization of the worst (maximum) future failures of the system. Besides the date t and the level x_t, the manager must specify the forecasted inflow $\hat{a}_t$ for the current day (i.e. the inflow into the lake during the next 24 hours). The forecast $\hat{a}_t$ can be a direct guess of the manager or it can be computed by the DSS itself through a formal mathematical forecasting model (see next section). In any case the suggestion of the DSS is an interval

$$R_t = [r'(t, x, \hat{a}_t), r''(t, x, \hat{a}_t)]$$

of possible releases. In other words, the operating rule, called min-max operating rule is not a single value function. The interval R_t reduces to a single release when the lake is too full or too empty, i.e. when the situation is critical for floods or agriculture.

If the forecast is perfect ($\hat{a}_t = a_t$) and $r_t \in R_t$, for any t, the indicators A and F at the end of the current year and the years to follow are guaranteed to be smaller than two preassigned values A' and F'. This guarantee holds if the future inflows are formed by any concatenation of K one-year-long "reference" inflow sequences. In the case of Lake Como K was equal to 7, and the reference inflow sequences were

those corresponding to the seven years of the period (1946-1981) that
the manager felt to be most critical from the hydrological point of
view. Among them we have 1976 and 1977 which were the most dry and wet
years of the last decades. The historical agricultural water deficit in
1976 was A=850 $10^6 m^3$ while the number of days of flood in 1977 was
F=42 d. In order to obtain an improvement the two preassigned maximum
allowable values A' and F' were fixed to 600 $10^6 m^3$ and 20 d,
respectively.

The two functions $r'(t, x, \hat{a}_t)$ and $r''(t, x, \hat{a}_t)$ can be determined
by off-line solving of a deterministic optimal control problem defined
over an infinite time horizon. The interested reader may refer to [10]
for the most general formulation of this min-max control problem. In the
same paper it is shown that by imposing a suitable "periodicity"
constraint, the solution of the problem can be obtained by solving a
mathematical programming problem defined over a finite horizon (the
year). Moreover, this mathematical programming problem can be decomposed
into 365 very simple subproblems (one for each day t), which in turn can
be solved analytically in the special case of linear reservoirs (see
[11]), or by applying a standard one-dimensional searching procedure
(see [12]) in the general case.

The application of the risk-averse approach to the case of Lake
Como is described in [13]. The main result of the analysis is that the
reduction of the maximum number of days of flood from 42 to 20 days and
of the maximum yearly deficit from 850 to 600 million cubic meters can
be guaranteed even when the forecast is done by means of real predictors
like those described in the next section. Moreover, the improvements one
can obtain with better predictors are only marginal. Actually, if the
release r_t is systematically selected in the upper range of the interval
R_t. i.e. if the manager is strongly in favour of flood protection, the
results obtained with real predictors are slightly better than those
obtained with the perfect forecast $\hat{a}_t = a_t$. This little surprise can be
explained as follows. In general, predictors underestimate the sudden
variations of the inflows during flood episodes. Thus, during these
periods the release is smaller than the one that would be computed with
a perfect predictor, the difference being a surplus of storage that can
be used later to satisfy the demand of the downstream users. The fact
that the performance of the system is rather insensitive to the reliability of the forecast is certainly a peculiar feature of Lake Como and
not of the min-max method. The reason is that in the case of Lake Como,
the action of the manager is rather limited during the floods. In fact,
as soon as the level of the lake is above the upper limit x" of the
control range, the manager has no option and must open the gates of the
dam completely. Thus, in these periods errors in the forecast do not
influence the final decision, so that the overall performance of the
system becomes rather insensitive to the forecasting technique.

The analysis of the risk-averse solution revealed that the interval
R_t shrinks during the high inflow seasons and at the end of the dry
season. During the rest of the year the interval R_t is rather large,
unless the level of the lake or the forecasted inflow are extremely high
or low. This means that most of the times, the manager can choose the

release which better fits the additional information he has, or he can accommodate for secondary objectives which were not considered in the formulation of the problem. For example, if the efficient or the generalized operating rules described in Sections 4 and 5 suggest a release $r_t \in R_t$, the manager can follow this suggestion without hesitating because he knows that under these circumstances minimization of the expected values of the indicator is not in conflict with the min-max criterion. However, if the release r_t suggested by the efficient (or generalized) operating rule does not belong to R_t, then the manager has to make a real trade-off between risk-aversion and mean-profit maximization. This is why the possibility of simulating the future behaviour of the lake (see next section) may be helpful.

7. SIMULATION AND FORECASTING

Simulating the future behaviour of the lake with the DSS is a very simple task. All one has to do is to specify which of the three operating rules $r(\cdot)$ available in the DSS one likes to use in the continuity equation (2). Therefore, all that is needed is to type in the inflow values for the next few days (in practice the simulation is never performed for more than 7 days, since the effect of the present release on the lake dynamics becomes negligible after such a period). The manager may choose the inflow values from his data bank by quickly looking up historical floods and droughts (see next section) or use any sequence of inflows he feels as particularly critical. But he also has the option of using the DSS to compute the most probable sequence of future inflows.

In fact, three different schemes have been developed for forecasting the inflows into Lake Como and implemented on the DSS. All of them are autoregressive forecasting models with exogenous inputs, i.e. the forecasted inflow is computed from the inflow in the preceding days and from the hydrological variables y_t^1 described in Section 5 (snow cover, aquifer depth, rainfall). They differ from each other in forecasting horizon, type of information used, and method employed to calibrate the parameters.

For the one-day-ahead forecast, the predictor that was finally implemented is described by the following equation.

$$\hat{a}_t = \alpha_t a_{t-1} + \beta_t a_{t-2} + \gamma_t y_t^2 + \delta_t y_t^3$$

where $\hat{a}_t$ is the inflow forecasted at the beginning of day t for the next 24 hours, α_t, β_t, γ_t, and δ_t are parameters, a_{t-1} and a_{t-2} are the inflow values in the past two days, y_t^2 is the depth of the aquifer at the beginning of day t, and y_t^3 is the total rainfall during days (t-1) and (t-2). The parameters are periodic over the year since they are somehow adapted to the seasonal variations of soil moisture, vegetation cover, evapotranspiration, etc., which regulate the rainfall-runoff process of the catchment. The parameters γ_t and δ_t are forced to be equal to zero when the inflow a_{t-1} in the preceding day is smaller than its mean plus one standard deviation. This means that when inflow are not too high only the autoregressive part of the model is used to forecast the inflow. However, during periods of high inflow, the exogenous inputs are also used. The parameters $(\alpha_t, \ldots, \delta_t)$ have been computed off-line with the least squares method and the performance of the predictor turns out to be quite satisfactory. The noise to signal ratio is 0.32 and the correlation between real and forecasted values is 0.95. If only the periods of fast inflow variations are considered (beginning of the floods), the above mentioned ratio becomes 0.40 and the correlation is still 0.91. However, as is common with all types of autoregressive predictors, peaks values are often underestimated. Nevertheless, this predictor represents a very useful tool for the manager, because the peak value of the flood does not influence his daily decision after all (see Section 6).

The second predictor implemented in the DSS is a linear forecasting model which determines the total inflow during the following three days on the basis of the individual values of the inflow in the three preceding days. The performance of this predictor is worse than that of the one-day-ahead predictor because the forecast horizon is longer. Nevertheless, the accuracy of the forecast has been partially improved by using the one day ahead prediction as a corrective term. The final overall performance, again measured as noise to signal ratio, is 0.62, with a slightly higher value (0.67) at the beginning of the floods.

The third predictor is a simple first-order autoregressive model for the 7 days ahead total inflow. The pattern of inflows over a 7 days period would seem relevant to the manager; however, a forecast of the total inflow seemed the only reasonable request for such a time span. The performance of this forecasting model is poorer than those of the preceding models, giving a correlation between forecasted and real values of 0.69 and a noise to signal ratio equal to 0.75.

8. DATA ANALYSIS AND RETRIEVAL

The basic operation performed by the DSS is data analysis and retrieval. Data transmitted by the telemetering network are checked for consistency and in the case the test fails, the manager is asked to decide if they should be rejected or not. All the daily data are stored in a master data bank, while whenever the manager feels they are of interest, also two-hours data are saved in a separate file, mainly containing flood episodes. Standard statistics are automatically computed every ten days and every month and compared with the long term values based on the last twenty years. The same kind of statistics are available at any time on request. At the end of each year, complete tables with daily values, the days, monthly and yearly statistics are printed in a form suitable for publication in the bulletin issued by the agency in charge of the regulation of the lake.

9. CONCLUSION

In its present configuration, the DSS performs a series of operation which help the manager to better evaluate the consequences of his daily decision, namely, the release from the lake. Three different rules for computing the release are available: Two of them attempt to maximize the average long term benefits of the system, while the third one is aimed at avoiding dramatic failures during extremely critical hydrological episodes. The DSS can also forecast the inflows into the lake, which in turn, can be used to simulate the lake dynamics in the near future. Finally, the DSS also performs standard statistical analysis of the collected data and prints out summary tables.

Many of the above operations are nothing but the optimal solutions of complex optimization problems, among which there are two stochastic multiobjective optimal control problems (Sections 4 and 5) and a min-max optimal control problem (Section 6). The optimal solutions of these problems have been obtained by off-line nonlinear mathematical programming, which has also been used to calibrate the three different forecasting models implemented in the DSS.

The manager has been using the DSS since the end of 1981. A detailed statistical analysis carried out on the data of the first years of operation [14] has shown that the DSS has positively influenced the manager who has, in particular, controlled the floods much more effectively than in the past. In the years that followed, the manager has sometimes significantly deviated from the suggestions of the DSS, but this is obviously due to the differences between the real-word circumstances and the assumptions on which the studies wefe based. For instance, the shutdown of a turbine for maintenance, or the temporary closure of and uptake point for repavement of an irrigation canal, may require releases very different from those suggested by the DSS. This fact is not a surprise, however, since the purpose of the DSS is not to substitute the manager, but rather to provide him with quick and easy tools he could use to get insights into the most probable consequences of his decision and into the practical value of all the hydrometeorological information which is made available to him in real time.

REFERENCES

[1] Guariso G., and Werthner H.: Environmental Decision Support Systems, Ellis Horwood, Chichester, UK 1989.
[2] Loucks D. P.(ed): Systems Analysis for Water Resources Management: Closing the Gap between Theory and Practice, IAHS Publications n. 180, IAHS Press, Wallingford, UK 1989.
[3] Fedra, K. (ed.): Expert Systems for Integrated Develpment: a Case Study of Shanxi Province The People's Republic of China. Final Report Voll. I,II,III. International Institute for Applied Systems

Analysis, A-2361 Laxenburg, Austria 1988.

[4] Zielinski P., Guariso G., and Rinaldi S.: A heuristic approach for improving reservoir management: Application to Lake Como, Proceedings International Symposium on Real-time Operation of Hydrosystems, Waterloo, Ontario, Canada, 1981.

[5] Guariso G., Rinaldi S., Soncini-Sessa R.:The Management of Lake Como: A Multiobjective Analysis, Water Resources Research, 22, 2, (1986), 109-120.

[6] Helweg O.J., Hinks R.W., and Ford D.T.: Reservoir system optimization, Water Resources Planning and Management Division of the ASCE, 108 (2) (1982), 169-179.

[7] Garofalo F., Raffa U., and Soncini-Sessa R.: Identification of Lake Como management policy (in Italian), Proceedings of the 17th Italian Symposium on Hydraulics, Palermo, Italy, 1980.

[8] Guariso G., Rinaldi S., and Soncini-Sessa R.: Analysis of the reliability of a proposal for the management of Lake Como (in Italian), Proceedings of the 18th Italian Symposium on Hydraulics, Bologna, Italy, 1982.

[9] Guariso G., Rinaldi S., and Zielinski P.: The value of information in reservoir management, Applied Mathematics and Computation, 15(2) (1984), 165-184.

[10] Orlovski S., Rinaldi S., and Soncini-Sessa R.: A min-max approach to storage control problems. Applied Mathematics and Computation 12 (1983), 237-254.

[11] Orlovski S., Rinaldi S., and Soncini-Sessa R.: A min-max approach to reservoir management. Water Resources Research 20(11) (1984), 1506-1514.

[12] Guariso G., Orlovski S., Rinaldi S., and Soncini-Sessa R.: A risk-averse approach the management of Lake Como, Proceedings of the 18th IFAC World Congress, Budapest, Hungary, 1984.

[13] Guariso G., Orlovski S., Rinaldi S., and Soncini-Sessa R.: An application of the risk-averse approach the management of Lake Como, Journal of Applied Systems Analysis 5(2) (1984), 54-64.

[14] Guariso G., Rinaldi S.,Soncini-Sessa R.,"Can a Microcomputer Help the Manager of a Multipurpose Reservoir? The Experience of Lake Como", Proc. Int. Symp. on Lake and Reservoir Management, Knoxville, Tennessee, USA, 1983, 575-579.

A DECISION GENERATOR SHELL IN PROLOG

K.M. van Hee, W.P.M. Nuijten
Eindhoven University of Technology, Eindhoven, The Netherlands

ABSTRACT

Many decision problems can be considered as searching for an element in a finite set. Classical approaches lead to sophisticated combinatorial optimization algorithms that exploit a lot of the structure of the decision situation. Often these algorithms are not easy to adapt to new constraints on decisions imposed by the decision makers.

Our approach to the generation of decisions is to use general search methods that are easy to adapt in case of new constraints. In general they give less good decisions but are more robust. A general-purpose shell based on these search methods is sketched using Prolog. As an illustration two decision problems are treated: the travelling salesman problem and precedence constrained scheduling.

1. Introduction

We use a well-known paradigm of systems theory cf. Checkland (1981) where there are two communicating systems, one to be controlled, called target or object system, and a control system, often called the information system. The target system sends status information to the information system and the information system sends decisions to the target system. The term 'decision' is used for a unit of information. A decision might be a simple control action that influences the target system for a short period of time, or a set or sequence of such actions that controls the target system for a longer period of time. An example of the last case is a production schedule or a route plan. There are at least two protocols for the communication between target and information system: one where the information system sends periodically new decisions to the target system and one where the information system only sends new decisions if the observed state of the target system satisfies some condition. It is clear that in general two successive decisions are dependent. For instance if the decision is a production schedule the next decision might be an adaptation of the foregoing one. Although decision making is a process, we consider only the production of one decision.

A Decision Support System (DSS) is an automated part of an information system that assists users in making decisions by:
- performing data management functions,
- evaluating the effects of user proposed decisions,
- generating decisions that satisfy some user defined conditions.

The evaluation function can be divided into two parts: verification whether the user proposed decision is allowed and a computation of quality measures for allowable decisions.

The generation function may also be split into several parts varying from stepwise development of a decision, where the user may choose in each step from a finite set of computed alternatives, to a fully automatically computed decision.

Our definition of a DSS is more restrictive than others, cf. for instance Keen and Scott Morton (1978). As advocated in Eiben and van Hee (1990) a DSS should be developed in the evolutionary style. This means that first the data management functions are developed, then the verification function, then the other evaluation functions, then the stepwise generation support and finally the automatic generation of decisions. In this approach the user's experience with the system evolves simultaneously with the functionality of the system.

In this paper we only consider the part of a DSS that generates decisions: the Decision Generator (DG). Many DSS's for operational planning have a DG based on sophisticated algorithms that rely on specific properties of the decision situation. For instance the simplex algorithm requires the object function and all constraints to be linear, algorithms for job shop scheduling may require that for each task there is only one possible machine available etc.. In many practical situations however the requirements of these algorithms are often not satisfied. Even if they are satisfied at the time the DSS was developed, they may be unsatisfied later due to changes in the decision situation. For DSS's for tactical or strategic decision making this phenomenon is not so severe. In these cases a DSS often considers a simplification of the real situation and the proposed decisions are of a rather global nature such as the capacity of storage space or the number of vehicles. In operational decision making however it is unacceptable that a DSS suggests an unexecutable decision or a decision that does not exploit all possibilities the decision maker has.

Although specialized algorithms may find optimal solutions in case all requirements are fulfilled we advocate to use a more robust approach that sometimes gives less good decisions but that guarantees the proposed decisions to be executable. In Van Hee and Lapinski (1989) an abstract machine for decision making that can easily be adapted to new restrictions in the decision situation is presented. In Eiben and van Hee (1990) a method for obtaining robust DG's based on graph search methods is introduced. In this paper we also consider DG's based on search methods. We generalize aforementioned method a bit and describe how it can be implemented in Prolog to obtain a DG shell. The shell can be used to create specific DG's for specific situations. If the situation changes one can easily generate a modified DG.

In the shell there is generic knowledge: this knowledge or code will be part of all systems composed with the shell. This part has to be defined by the designer of the shell. Then there is a part of the knowledge that is fixed for a problem type, this is called <u>domain knowledge</u>. This part has to be defined by the DG-expert. The third part of knowledge is dependable on the instance of the decision situation and can be defined by the user or decision maker (<u>instance knowledge</u>). This part is not restricted to factual knowledge as usually is the case but it also concerns rules such as restrictions on decisions and goals to be met.

The paper is organized as follows. In section 2 a method for the construction of DG's is sketched. In section 3 a Prolog implementation of the generic knowledge of the shell is given. In section 4 we consider two problem types to illustrate how the domain knowledge can be defined. Finally in section 5 we sketch the architecture of the shell.

2. Constructing a decision generator

In this section a method for the construction of DG's is sketched. As mentioned before we only consider DG's based on search methods.

As stated in section 1 a DG should generate a decision satisfying some conditions. Such a decisions will be called a *solution*. As every decision is a candidate for being a solution we refer to every decision as a *candidate*. The first step to be made by a DG-expert is defining what the set of candidates is, i.e. the set in which a solution is searched. Searching is performed on the basis of a row of candidates. This list is called the *search state* of the search process. The next step is the definition of a binary relation on the candidates. This relation is called the neighbourhood relation. Only candidates that are neighbours of candidates already in the search state are considered. So our search methods are local search methods (cf. Papadimitriou and Steiglitz (1982)). Requirements for a neighbourhood could be:
- a neighbourhood should not be too large,
- a neighbourhood should be easy to generate,
- from a starting point every candidate should be reachable through the neighbourhood relation.

Next a goal is defined, i.e. a predicate on the candidates defining which candidates are solutions. Also some additional constraints on candidates can be defined. Formally these additional constraints are represented by a predicate *feasible* on the candidates (a candidate satisfying the constraints is called feasible). These goal- and feasibility predicates should be definable by the user (instance knowledge). The feasibility predicate is used to filter the neighbours and eliminating all non-feasible neighbours. So in effect by using the feasibility predicate search can be restricted to any subset of the set of candidates. We remark that the use of a feasibility predicate often is recommended as in general the set of candidates is far too large. As the feasibility predicate is easily redefined it gives our search methods a great deal of flexibility.

A *search step* is a transition of the current search state to a new search state. This is done by:
- selecting a candidate p from the current search state,
- generating all neighbours of p,
- eliminating all non-feasible neighbours,
- updating the search state from the old search state and the feasible neighbours, i.e. determining which candidates are to be maintained in the new search state and in what order.

The first step is standardized: always select the first candidate.

A search method is an iteration process of search steps. The search process continues until a solution is found. Formally this can be described as follows.

Let C be the set of candidates, $N \subseteq C \times C$ the neighbourhood relation, $S = C^*$ (the set of all finite rows from C) the search state space and $T : S \longrightarrow S$ the transition function (defining the search steps). We remark that for the sake of convenience we sometimes treat elements of S (rows) as sets. With $s \in S$ it holds that $T(s)$ is a permutation of a subset of $s \cup \{c \in C \mid \exists d \in s : (d,c) \in N \land \text{feasible}(c) \}$, i.e. $T(s)$ is a permutation of a subset of all candidates in search state s joined with all feasible neighbours of all candidates in s. A search method can be described by a search function $F : S \longrightarrow S$ which is defined as follows, with $s \in S$:

$$F(s) = F(T(s)) \qquad \text{if } \forall p \in s : \neg \text{ solution}(p),$$
$$F(s) = s \qquad\qquad \text{if } \exists p \in s : \text{solution}(p).$$

Let A be defined by $A = \{ s \in S \mid \exists p \in s : \text{solution}(p) \}$, let $T^n(s)$ be defined by $T^n(s) = T(T^{n-1}(s))$ and $T^0(s) = s$ for $n \in \mathbb{N}$, and let s_0 be some starting search state in S.

In case for all $n \in \mathbb{N}$, $T^n(s_0) \notin A$ holds, $F(s_0)$ is undefined. In that case, since S is finite, the sequence $T^n(s_0)$ has to be periodically, i.e. for some $p \in \mathbb{N}$ and all $n \in \mathbb{N}$:

$$T^{n+p}(s_0) = T^n(s_0).$$

However if F is defined for s_0 then $F(s_0)$ is totally specified by T.

3. A Prolog implementation of the generic knowledge

In this section a Prolog implementation of the generic knowledge is given. A background knowledge of Prolog is therefore presumed (cf. Schnupp and Bernhard (1986)).

As stated before we are developing DG's based on search methods. A component of a search method is the search strategy. At the end of this section some examples of search strategies are given. By treating the generic knowledge, components of the domain knowledge and the instance knowledge will be encountered.

We state that in general a search method consists of an initialization step followed by zero or more search steps. We define the *state* of a search method to be a list of candidates that could be used for further searching (see section 2). The foregoing is reflected in the Prolog-rule:

```
solve : -
    initialize(State!),
    search_loop(State?).
```

The ? (input) and ! (output) notation is borrowed from the specification language Z (cf. Spivey (1989)). Suffixing a variable with ? means that the variable needs to have a value when the predicate is invoked. Suffixing a variable with ! means that if all input variables have a value a new value for this output variable is calculated. The notation has no further meaning: State? and State! both address the same variable.

The predicate *initialize* produces an initial list of candidates and is to be defined by the user (instance knowledge).
The predicate *search_loop* performs zero or more search steps. Searching is stopped if at least one solution is encountered in the present search state. If so all solutions in the search state are displayed. If no solution is encountered a search step is performed yielding a new search state and the search process continues with that new search state.

```
search_loop(State?) : -
    goal(State?),
    display_solutions(State?).

search_loop(State?) : -
    search_step(State?, NewState!),
    search_loop(NewState?).
```

Notice that:
- *goal* checks whether there is a solution in *State*,
- *display_solutions* displays all solutions in *State*.

The predicates *goal* and *display_solutions* use the predicates *solution* and *display*. The predicate *solution* defines which candidates are solutions by checking whether the value of some evaluation function on the candidates is below a certain boundary. Without loss of generality we assume the better a candidate is the less its value is. Observe that a special form of the *goal* predicate is used here. We state the general format for *solution* to be:

```
solution(Candidate?) : -
    eval(Candidate?, Value!),
    Value <= boundary.
```

where
- *eval* is a user defined evaluation function on the candidates,
- boundary is a user defined constant.
The predicate *display* displays a candidate and is to be defined by the DG-expert (domain knowledge).

```
goal([Cand | Cands]?) : -
    solution(Cand?).

goal([Cand | Cands]?) : -
    goal(Cands?).
```

display_solutions([]).

display_solutions([Cand | Cands]?) : -
    solution(Cand?),
    display(Cand?),
    display_solutions(Cands?).

display_solutions([Cand | Cands]?) : -
    display_solutions(Cands?).

We remark that the *goal* predicate can be expanded with a parameter valued with the current time. Herewith we enable the user to control the computation time of the search process. We suggest that after the search process is stopped because the time limit was exceeded the user decides whether to continue the search process and if so may update the current search state (e.g. delete, change or produce new candidates). In that way an interactive DG is yielded. For the sake of convenience we omit the elaboration of this idea.

The predicate *search_step* calculates a new search state form the old one by
- selecting the first candidate in the present search state for further search,
- generating all neighbours of the selected candidate by means of the predicate *generate* to be defined by the DG-expert (domain knowledge),
- removing all non-feasible neighbours by means of the predicate *filter*,
- finally calculating a new search state from the feasible neighbours and the old search state by means of the predicate *update*.

    search_step([Cand | State]?, NewState!) : -
        generate (Cand?, Neighbours!),
        filter (Neighbours?, FeasibleNeighbours!),
        update (FeasibleNeighbours?, [Cand | State]?, NewState!).

Note that both *generate* and *filter* may produce an empty list of candidates.

The predicate *filter* uses the user defined predicate *feasible* which defines the feasible candidates (instance knowledge). This gives the problem solving method its flexibility (see section 2).

    filter([],[]).

    filter([X | Y]?, [X | Z]!) : -
        feasible(X?),
        filter(Y?, Z!).

    filter([X | Y]?, Z!) : -
        filter(Y?, Z!).

The predicate *update* defines the search strategy. There is a library of '*update*-predicates' all defining a search strategy of which the user may choose one (see end of this section).

In conclusion the following components are to be defined by the DG-expert (domain knowledge):
- a neighbourhood relation of candidates: *generate*.
- a procedure for presenting a solution: *display*.
The user has the following components to control the search process:
- a procedure for making an initial set of candidates: *initialize*,
- a predicate defining the feasible candidates: *feasible*,
- an evaluation function *eval* on candidates together with a boundary,
- a search strategy (an *update* predicate).

Above made partitioning is arbitrary. By shifting more components to be defined by the DG-expert the level of abstraction is enhanced. By shifting more components to be defined by the user flexibility is gained, but the user has to have more domain knowledge.
A so far untreated aspect of instance knowledge is background data. This data is used in the other predicates. We here state that the DG-expert should define a format for all data and possibly provide a tool for data manipulation.

Next we define five '*update* predicates' each describing a search strategy (cf. Pearl (1971) for the first four strategies). An *update* predicate determines which candidates from the current search state and the feasible neighbours are to be maintained in the next search state and in what order.

Depth-first search

By appending all feasible neighbours to the front end of the old search state after deleting the first candidate and taking the result as the new search state depth-first search is yielded. The predicate *append* is a system predicate putting its first argument in front of its second argument obtaining its last argument.

```
update(FeasibleNeighbours?, [Cand | State]?, NewState!) :-
    append(FeasibleNeighbours?, State?, NewState!).
```

Observe that as the list of feasible neighbours can be empty backtracking is enabled.

Breadth-first search

By appending all feasible neighbours to the end of the old search state after deleting the first candidate and taking the result as the new search state breadth-first search is yielded.

```
update(FeasibleNeighbours?, [Cand | State]?, NewState!) :-
    append(State?, FeasibleNeighbours?, NewState!).
```

Best-first search

By first appending all feasible neighbours to the old search state after deleting the first candidate and then sorting all candidates in the resulting search state best-first search is yielded. Sorting is done according to the *eval* predicate: candidates with the lowest 'eval-values' (the best candidates) are placed at the beginning of the search state.

    update(FeasibleNeighbours?, [Cand | State]?, NewState!) : -
        append(FeasibleNeighbours?, State?, TempState!),
        sort(TempState?, NewState!).

Hill climbing

By first sorting all feasible neighbours and then appending the sorted neighbours to the front end of the old search state after deleting the first candidate hill climbing is yielded. Here sorting is also done according to the *eval* predicate (see best-first search).

    update(FeasibleNeighbours?, [Cand | State]?, NewState!) : -
        sort(FeasibleNeighbours?, SortedNeighbours!),
        append(SortedNeighbours?, State?, NewState!).

Simulated Annealing

The following definition of *update* yields simulated annealing cf. Aarts and Korst (1989). Simulated annealing is a randomized search strategy where the search state always contains exactly one candidate. A neighbour is generated randomly . Let *valn* be the value of this neighbour and *valc* the value of the original candidate. Whenever the neighbour is not worse ($valn \leq valc$) it is accepted as the next candidate with which the search is continued. If the neighbour is worse ($valn > valc$) it is accepted with chance

$$\exp \left(\frac{valc - valn}{c} \right)$$

where $c \in \mathbb{R}^+$ is a constant. If the neighbour is not accepted the process continues by again randomly generating a neighbour etc.

Here the *update* predicate is an invokement of the predicate *sima*.

    update(FeasibleNeighbours?, [Cand]?, [NewCand]!) : -
        sima(FeasibleNeighbours?, Cand?, seed?, NewCand!).

where *seed* is a randomly obtained integer (we assume the availability of a random number generator) used for the randomization of simulated annealing.
The predicate *sima* uses the predicates:
- *length(A,L)* where L becomes the length of list A (a system predicate),
- *take(P,L,E)* where E becomes the element in list L on place P.
- *sima2(F,C,Vc,N,Vn,S,New)* where *New* becomes N if N is accepted (this is decided on the basis of Vc, Vn and S). If N is not accepted a new invokement of *sima* results.

Observe that the in *sima* used functions *exp* and / are yet to be defined. This assumes the availability of real arithmetic. There are Prolog versions with real arithmetic. If one does not have such a version, real arithmetic can be simulated by integers. We remark that a, b, c en m are constants.

```
sima(FeasibleNeighbours?, Cand?, Seed?, NewCand!) :-
    length(FeasibleNeighbours?, L!),
    Place is ((a * Seed + b) mod L) + 1,
    take(Place?, FeasibleNeighbours?, Neigh!),
    eval(Cand?, Vc!),
    eval(Neigh?, Vn!),
    NextSeed is (a * Seed + b) mod m,
    sima2(FeasibleNeighbours?, Cand?, Vc?, Neigh?, Vn?, NextSeed?, NewCand!).

sima2(FeasibleNeighbours?, Cand?, Vc?, New?, Vn?, Seed?, New!):-
    exp((Vc - Vn) / c) >= Seed / m.

sima2(FeasibleNeighbours?, Cand?, Vc?, Neigh?, Vn?, Seed?, New!).
    NewSeed is (a * Seed + b) mod m,
    sima(FeasibleNeighbours?, Cand?, NewSeed?, New!).

take(1, [Cand | Cands]?, Cand!).

take(Place?, [Cand | Cands]?, CandRes!) :-
    Place > 1,
    PlaceMin1 is Place - 1,
    take(PlaceMin1?, Cands?, CandRes!).
```

4. Examples of domain and instance knowledge

In this section the predicates that (in section 3) were left to be defined by the DG-expert and the user are defined for the Travelling Salesman Problem (TSP) cf. Lawler, Lenstra, Rinnooy Kan and Shmoys (1985) and Precedence Constrained Scheduling (PCS) cf. Garey and Johnson (1978).

4.1 The Travelling Salesman Problem

We start with a mathematical description of the problem type. Let
- C be a finite set of cities and $N = |C|$.
- dist : $C \times C \longrightarrow \mathbb{R}_0^+$ be a function denoting the distance between cities.

A <u>route</u> is a bijection r : $\{1,..,N\} \longrightarrow C$: a permutation of all cities. Let R be the set of all routes.

The function length : $R \longrightarrow \mathbb{R}$ is defined for every $r \in R$:

$$length(r) = \sum_{i=1}^{N-1} dist(r(1),r(i+1)) + dist(r(N),r(1))$$

denoting the length of a route.

A route with its length below a certain boundary is what we are looking for.

In Prolog the function *dist* is represented by a set of facts *dist(c1,c2,d)* where *d* is the distance from *c1* to *c2*.
The next question is the representation of routes in Prolog. We decide to represent a route by a list.
The neighbourhood of a route is chosen to be all routes generated by switching two adjacent cities (with exception of switching the first and the last city). Remark that we assume a route to be given. This implies that initially at least one route must be present in the search state. The predicate *generate* uses the predicate *hgenerate*. The latter has two input arguments:
- the first is a begin part of the original route,
- the second is a end part of the original route.
It holds that by appending the begin part at the front end of the end part the original route is yielded. A neighbour is produced by switching the first two cities of the end part and then appending the begin part at the front end of the thus generated list. Furthermore the first city of the end part is appended at the end of the begin part yielding the new begin part. The same city is then deleted from the end part.

```
generate(Route?, Neighbours!) :-
    hgenerate([],Route?,Neighbours!).

hgenerate(_,[City]?,[]).

hgenerate(BeginPart?, [City1, City2 | EndPart]?, [Neighbour | Res]!) :-
    append(BeginPart?, [City2, City1 | EndPart]?, Neighbour!),
    append(BeginPart?, [City1]?, BeginPartNew!),
    hgenerate(BeginPartNew?, [City2 | EndPart]?, Res!).
```

We remark that this is not the only possible neighbourhood or even the best possible. Another example of a TSP neighbourhood is one containing all 2-changes of cities and not just the 2-changes of adjacent cities.

After this 'expert part' (domain knowledge) the components to be defined by the user are treated. We evaluate a route by its length (the predicate *last* picks the last element of a list):

```
eval([First | Tail]?, Value!) :-
    last(Tail?, Last!),
    dist(Last?, First?, X!),
    heval([First | Tail]?, Dist!),
    Value is Dist + X.
```

```
heval([City]?, 0!).

heval([City1, City2 | Tail]?, Dist!) :-
    heval([City2 | Tail]?, DistRest!).
    dist(City1?, City2?, X!),
    Dist is DistRest + X,
```

Any *initialize* predicate yielding a route will do.
The user can state that all routes are feasible. This would result in the following
definition of *feasible*:

```
feasible(Route?).
```

Of course any of the treated search strategies can be chosen.

4.2 Precedence Constrained Scheduling

Again we start with a mathematical description. Let
- M be a finite set of machines,
- J be a finite set of jobs,
- able $\subseteq$ M $\times$ J such that (m,j) $\in$ able means machine m can perform job j,
- pre $\subseteq$ J $\times$ J such that (j',j) $\in$ pre denotes that j' should be completed before j is started
  (j' is a predecessor of j),
- dur : able $\longrightarrow$ $\mathbb{R}_0^+$ such that dur(m,j) denotes the amount of time needed by machine m
  to perform job j.

In a schedule jobs are attributed by machines and beginning times, that is a <u>schedule</u> is
a pair (m,b) of partial functions, where
- dom(b) = dom(m),
- m : J $\rightarrowtail$ M is such that $\forall$ j $\in$ dom(m) : (m(j),j) $\in$ able,
- b : J $\rightarrowtail$ $\mathbb{R}_0^+$ is such that
  - all jobs have their predecessors ready:
    $\forall$ j $\in$ dom(b) $\forall$ j' $\in$ J : (j',j) $\in$ pre $\Rightarrow$ j' $\in$ dom(b) $\land$ b(j') + dur(m(j'),j') $\leq$ b(j),
  - the processing of two jobs on the same machine cannot overlap:
    $\forall$ j,j' $\in$ dom(b) :
    m(j) = m(j') $\land$ j $\neq$ j' $\Rightarrow$ b(j) $\geq$ b(j') + dur(m(j'),j') $\lor$ b(j') $\geq$ b(j) + dur(m(j),j)

Let S be the set of all schedules.

The function c : S $\longrightarrow$ $\mathbb{R}_0^+$ is defined for every (m,b) $\in$ S:

$$c((m,b)) = \max \{ b(j) + dur(m(j),j) \mid j \in dom(m) \},$$

denoting the completion time of a schedule.

A <u>complete</u> schedule is a schedule (m, b) ∈ S where all jobs are scheduled, i.e. dom(m) = dom(b) = J. A complete schedule with its completion time under a certain boundary is what we are looking for.

In Prolog the set of jobs is represented by the fact *jobs(joblist)* where *joblist* is a list of all jobs. The set of machines is represented by the fact *machines(machinelist)* where *machinelist* is a list of all machines. The set *able* is represented by a set of facts *able(job, machinelist)* where *machinelist* is a list of all machines on which *job* can be performed. The set *pre* is represented by a set of facts *pre(job, preds)* where *preds* is a list of all predecessors of *job*. The function *dur* is represented by a set of facts *dur(mach, job, d)* where *d* is the duration of *job* on machine *mach*. All this is background knowledge.

We represent schedules by a list of operations. An operation is a term *operation(job, mach, bt, ct)* where *mach* is the machine on which *job* is performed, *bt* is the beginning time and *ct* the completion time of *job*.

We state a job on a machine only to be added to a schedule with its beginning time equal to the maximum of:
- the maximum completion time of any job on the same machine,
- the maximum completion time of any predecessor of the job.

The neighbourhood of a schedule is chosen to be those schedules generated by extending the schedule with one job on a machine that has the earliest possible beginning time among the beginning times of all job-machine combinations. Note that if more jobs can be started at the same time more neighbours will be generated. To generate all neighbours we first determine which unscheduled jobs have all predecessors scheduled. Next we determine which operations have the earliest beginning time. After that we construct the neighbours by appending all operations found to the original schedule each yielding a new schedule. All this is reflected in the following definition of *generate*:

```
generate(Schedule?, Neighbours!) :-
    all_preds_sched(Schedule?, Jobs!),
    earliest_operations(Schedule?, Jobs?, OperList!),
    make_neigh(Schedule?, OperList?, Neighbours!).
```

where
1) *all_preds_sched* determines which unscheduled jobs have all predecessors scheduled,
2) *earliest_operations* determines operations for all jobs found above which have the earliest possible beginning time,
3) *make_neigh* constructs the neighbours by appending all operations found to the original schedule each yielding a new schedule.

ad 1)

The predicate *all_preds_sched* uses the predicate *hall_preds_sched(S,J,N)* where *N* becomes a list of unscheduled jobs from job list *J* that have all predecessors scheduled in schedule *S* (see Appendix for the definition of *hall_preds_sched*).

```
all_preds_sched(Schedule?, NoPreds!) :-
    jobs(AllJobs!),
    hall_preds_sched(Schedule?, AllJobs?, NoPreds!).
```

ad 2)

The predicate *earliest_operations* uses the predicates
- *earliest_machs(S,J,M,O)* where *O* becomes a list of operations denoting on which machines from machine list *M* job *J* can be started as early as possible given schedule *S*.
- *new_operationlist(O1,O2,O3)* where *O3* becomes a list of the earliest possible operations in *O1* and *O2* (all operations in *O1* have the same beginning time and so do all operations in *O2*) (see Appendix for the definition of *new_operationlist*).

```
earliest_operations(Schedule?, [Job]?, OperList!) :-
    able(Job?, Machs!),
    earliest_machs(Schedule?, Job?, Machs?, OperList!).

earliest_operations(Schedule?, [Job | Jobs]?, OperList!) :-
    earliest_operations(Schedule?, Jobs?, OperList1!),
    earliest_operations(Schedule?, [Job]?, OperList2!),
    new_operationlist(OperList1?, OperList2?, OperList!).
```

The predicate *earliest_machs* uses the predicate *make_operation(S,J,M,O)* where *O* becomes a list of one operation *(J,M,BT,CT)* with
- *BT* the earliest possible beginning time after the last job on machine *M* and after the last predecessor of job *J*,
- *CT* is the completion time of job *J* (*CT = BT + dur(M,J)*).

```
earliest_machs(Schedule?, Job?, [], []).

earliest_machs(Schedule?, Job?, [Mach | Machs]?, OperList!) :-
    earliest_machs(Schedule?, Job?, Machs?, OperList1!),
    make_operation(Schedule?, Job?, Mach?, OperList2!),
    new_operationlist(OperList1?, OperList2?, OperList!).
```

The predicate *make_operation* uses the predicates
- *comptime_mach(S,M,T)* where *T* becomes the maximum completion time of any job on machine *M* in schedule *S* (see Appendix for the definition of *comptime_mach*),
- *comptime_preds(S,J,T)* where *T* becomes the maximum completion time of any predecessor of job *J* in schedule *S* (see Appendix for the definition of *comptime_preds*),
- *max(A,B,C)* where *C* becomes the maximum of *A* and *B* (not elaborated).

```
make_operation(Schedule?, Job?, Mach?, Oper!) : -
    comptime_mach(Schedule?, Mach?, Time1!),
    comptime_preds(Schedule?, Job?, Time2!),
    max(Time1?, Time2?, Begintime!),
    dur(Mach?, Job?, Dur!),
    Comptime is Begintime + Dur,
    Oper = [operation(Job, Mach, Begintime, Comptime)].
```

ad 3)

The predicate *make_neigh* is rather straightforward.

```
make_neigh(Schedule?, [], []).
```

```
make_neigh(Schedule?, [Oper | OperList]?, [[Oper | Schedule] | Neighbours]!) : -
    make_neigh(Schedule?, OperList?, Neighbours!).
```

As mentioned before a complete schedule with its completion time under a certain boundary is what we are looking for. We evaluate a schedule by its completion time plus a penalty. The penalty of a schedule is the sum of the maximum duration of all unscheduled jobs. Observe the penalty for a complete schedule to be equal to zero, so complete schedules are evaluated strictly by their completion time.

```
eval(Schedule?, Value!) : -
    penalty(Schedule?, Pen!),
    comptime(Schedule?, CT!),
    Value is Pen + CT.
```

The predicate *comptime(S,T)* calculates the completion time T of a schedule S (i.e. the maximum completion time of any scheduled job).

```
comptime([], 0).
```

```
comptime([operation(_,_,_,Time1) | OperList]?, Value!) : -
    comptime(OperList?, Time2!),
    max(Time1?, Time2?, Value!).
```

The penalty of a schedule is calculated by first determining which jobs are unscheduled. This is done by deleting the scheduled jobs from a list of all jobs. From the unscheduled jobs the sum of all maximum durations is calculated obtaining the penalty. The predicate *penalty(S,P)* uses the predicates
- *schjobs(S,SJ)* where *SJ* becomes a list of all scheduled jobs in schedule *S*.
- *unschjobs(AJ,SJ,UJ)* where *UJ* becomes a list of all jobs in job list *AJ* that are not in job list *SJ*.
- *totaldur(UJ,P)* where *P* becomes the sum of all the maximum durations of the jobs in *UJ*.
(see Appendix for the definition of *schjobs*, *unschjobs* and *totaldur*)

```
penalty(Schedule?, Pen!) :-
    schjobs(Schedule?, SchedJobs!),
    jobs(Alljobs!),
    unschjobs(Alljobs?, SchedJobs?, UnSchedJobs!),
    totaldur(UnSchedJobs?, Pen!).
```

We choose the following definition of the *initialize* predicate yielding a list containing just the empty schedule:

```
initialize([[]]).
```

Of course any of the treated search strategies can be chosen.

5. The DG shell

The DG shell is in fact a tool kit to construct domain specific DG's. These DG's can be used as a part of a DSS (see section 1). Up to now we sketched only a small part of the tool kit. The tool kit has the following components, besides a Prolog interpreter:
- **generic knowledge base**: containing the predicate definitions of section 3.
- **domain knowledge definition facility**: to define predicates like generate, display, eval and feasible. Furthermore this facility enables the expert to define the predicate names of the facts of the instance knowledge to be defined by the user. In addition to these predicate names the expert should define the arity of these predicates, the types of the parameters and possibly some predicates checking the consistency of the facts.
- **instance data entry facility**: to enter facts, using the above mentioned names, arities, types and consistency checks.
- **control facility**: to compose a specific search method by selection of a search strategy, an evaluation predicate and a feasibility predicate. Furthermore this facility should allow the user to change parameters during the search process. It depends on the expertise of the user if he is able to define evaluation or feasibility predicates. An unexperienced user should only select these predicates from the domain knowledge base. The predicate names of facts are considered as domain knowledge here. In case of the TSP only *dist* was such a predicate. In the case of PCS *jobs, machines, able, pre* and *dur* were such predicates. It is clear that only facts of the form *able(j, [m1,m2])* are allowed if *j* is a job and *m1* and *m2* are machines. Hence a consistency check to guarantee referential integrity is required here. In fact the shell should have all standard database management functions.

Appendix

In this appendix some predicates used in section 4 are elaborated.

The predicate *hall_preds_sched(S,J,N)* calculates a list *N* of unscheduled jobs from job list *J* that have all predecessors scheduled in schedule *S*. In the definition of *hall_preds_sched* the following predicates are used:
- *member(X,Y)* a system predicate checking whether *X* is an element of list *Y*,
- *allmember(X,Y)* checking whether *X* is a subset of *Y*.

```
hall_preds_sched(Schedule?, [], []).

hall_preds_sched(Schedule?, [Job | Jobs]?, NoPreds!) :-
    member(operation(Job,_,_,_), Schedule?),
    hall_preds_sched(Schedule?, Jobs?, NoPreds!).

hall_preds_sched(Schedule?, [Job | Jobs]?, [Job | NoPreds]!) :-
    not (member(operation(Job,_,_,_), Schedule?)),
    pre(Job?, Preds!),
    allmember(Preds?, Schedule?),
    hall_preds_sched(Schedule?, Jobs?, NoPreds!).

hall_preds_sched(Schedule?, [Job | Jobs]?, NoPreds!) :-
    not (member(operation(Job,_,_,_), Schedule?)),
    pre(Job?, Preds!),
    not (allmember(Preds?, Schedule?)),
    hall_preds_sched(Schedule?, Jobs?, NoPreds!).

allmember([], Schedule?).

allmember([Pred | Preds], Schedule?) :-
    member(operation(Pred,_,_,_), Schedule?),
    allmember(Preds?, Schedule?).
```

The predicate *new_operationlist(O1,O2,O3)* calculates which operations in operation lists *O1* and *O2* have the earliest beginning times resulting in *O3* and uses the predicate *time(O,T)* where *T* becomes the beginning time of the first operation in *O* (*O* not empty).

```
new_operationlist([], OperList?, OperList!).

new_operationlist(OperList?, [], OperList!).

new_operationlist(OperList1?, OperList2?, OperList1!) :-
    time(OperList1?, Time1!),
    time(OperList2?, Time2!),
    Time1 < Time2.
```

```
new_operationlist(OperList1?, OperList2?, OperList2!) : -
    time(OperList1?, Time1!),
    time(OperList2?, Time2!),
    Time1 > Time2.

new_operationlist(OperList1?, OperList2?, OperList!) : -
    time(OperList1?, Time1!),
    time(OperList2?, Time2!),
    Time1 = Time2,
    append(OperList1?, OperList2?, OperList!).

time([operation(_,_,T,_) | OperList], T).
```

The predicate *comptime_mach(S,M,T)* calculates the maximum completion time *T* of any job on machine *M* in schedule *S*.

```
comptime_mach([], Mach?, 0).

comptime_mach([operation(_,Mach1,_,_) | OperList]?, Mach2?, Time!) : -
    not (Mach1 = Mach2),
    comptime_mach(OperList?, Mach?, Time!).

comptime_mach([operation(_,Mach,_,Time1) | OperList]?, Mach?, Time!) : -
    comptime_mach(OperList?, Mach?, Time2!),
    max(Time1?, Time2?, Time!).
```

The predicate *comptime_preds(S,J,T)* calculates the maximum completion time *T* of any predecessor of job *J* in schedule *S*.

```
comptime_preds(Schedule?, Job?, Time!) : -
    pre(Job?, Preds!),
    hcomptime_preds(Schedule?, Preds?, Time!).

hcomptime_preds(Schedule?, [Pred]?, Time!) : -
    member(operation(Pred,_,_,Time), Schedule?).

hcomptime_preds(Schedule?, [Pred | Preds]?, Time!) : -
    hcomptime_preds(Schedule?, Preds?, Time1!),
    member(operation(Pred,_,_,Time2), Schedule?),
    max(Time1?, Time2?, Time!).
```

The predicate *schjobs(S,J)* calculates a list *J* of all scheduled jobs in schedule *S*.

```
schjobs([],[]).

schjobs([operation(Job,_,_,_) | Y]?, [Job? | X!]) : -
    schjobs(Y?, X!).
```

The predicate *unschjobs(AJ,SJ,UJ)* calculates a list *UJ* of unscheduled jobs from the list *AJ* of all jobs and a list of *SJ* of scheduled jobs.

```
unschjobs([], Schjobs?, [])

unschjobs([X | Y]?, Schjobs?, UnSchedJobs!) :-
    member(X?, Schjobs?),
    unschjobs(Y?, Schjobs?, UnSchedJobs!).

unschjobs([X | Y]?, Schjobs?, [X | UnSchedJobs]!) :-
    not(member(X?, Schjobs?)),
    unschjobs(Y?, Schjobs?, UnSchedJobs!).
```

The predicate *totaldur(JL,P)* calculates the sum *P* of all maximum durations of all jobs in job list *J* and uses the predicate *maxdur(J,T)* where *T* becomes the maximum duration of job *J* on any machine.

```
totaldur([], 0).

totaldur([Job | Jobs]?, Pen!) :-
    maxdur(Job?, M!),
    totaldur(Jobs?, Pen1!),
    Pen is M + Pen1.
```

The predicate *maxdur(J,M)* calculates the maximum duration *M* of job *J* and uses the predicates
- *durlist(J,ML,DL)* where *DL* becomes a list of the durations of job *J* on all machines in machinelist *M*,
- *maxlist(DL,M)* where *M* becomes the maximum in list *DL*.

```
maxdur(Job?, M!) :-
    able(Job?,MachList!),
    durlist(Job?,MachList?,DurList!),
    maxlist(DurList?, M!).

durlist(Job?, [], []).

durlist(Job?, [Mach | MachList]?, [Dur | DurList]!) :-
    dur(Mach?, Job?, Dur!),
    durlist(Job?, MachList?, DurList!).

maxlist([], 0).

maxlist([X | Y]?, Z!) :-
    maxlist(Y?, R!),
    max(X?, R?, Z!).
```

References

Aarts, E.H.L. and Korst, J., Simulated annealing and Boltzmann machines: a stochastic approach to combinatorial optimization and neural computing, Wiley, 1989.

Checkland, P., Systems Thinking and Systems Practice, Wiley, 1981.

Eiben, A.E. and van Hee, K.M.: Knowledge Representation and Search Methods for Decision Support Systems, in: Gaul, W. and Schader, M., Ed., Data, Expert Knowledge and Decisions, Springer-Verlag, 1990.

Garey, M.R. and Johnson, D.S., Computers and Intractability: A Guide to the Theory of NP-completeness, Freeman and Co, 1978.

van Hee, K.M. and Lapinski, A., OR and AI approaches to decision support, Decision Support Systems 4 (1989), pp 447-459.

Keen, P.G.W. and Scott Morton, M.S., Decision support systems: an organized perspective, Addison-Wesley, 1978.

Lawler, E.L., Lenstra, J.K., Rinnooy Kan, A.H.G. and Shmoys, D.B., The Traveling Salesman Problem: A Guided Tour of Combinatorial Optimization, Wiley, 1985.

Papadimitriou, C.H. and Stieglitz, K., Combinatorial optimization: algorithms and complexity, Prentice-Hall, 1982.

Pearl, J.R., Artificial intelligence: the heuristic programming approach, McGraw-Hill, 1971.

Schnupp, P. and Bernhard L.W., Productive Prolog Programming, Prentice Hall, 1986.

Spivey, M., The Z notation, Prentice Hall, 1989.

A MULTI OBJECTIVE DECISION SUPPORT SYSTEM FOR PUBLIC PLANNING

R. Janssen, M. van Herwijnen
Free University, Amsterdam, The Netherlands

ABSTRACT

Our decision support system DEFINITE is intended to support decisions with a finite set of alternatives in relation to a finite number of criteria. It can be used to generate, compare and evaluate alternatives as a preparation to policy decisions. The system intends to function as a learning, comminication and decision making tool for one or more decision makers.

1. INTRODUCTION

Decision support systems can be defined as: "Interactive computer-based systems that help decision makers utilize data and models to solve unstructured problems" [1]. Although this definition is now almost twenty years old, and although numerous competing definitions have been invented [2][3], this definition contains all elements that are essential to the concept of decision support.

Interactive and computer-based are essential feautres of DSS. By using a DSS decision makers no longer communicate with models through an analyst but have direct access to models and information that are available to support the decision. By bridging the gap between methods and decision maker Decision Support Systems induce a shift in the use of the methods from the analyst to the decision maker. **Help the decision makers** highlights the point that decision makers as individuals are supported. The actual decision making process as it takes place at the conference table is not supported. **Utilize data and models** emphasizes that support is given by processing information. Design of a decision support should reflect the individuals capacities to process information.

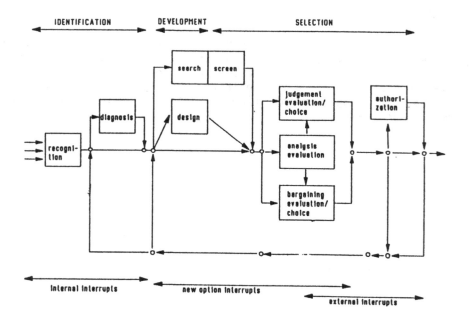

Figure 1. A general model of the strategic decision process [4].

Multiobjective Decision Support Systems (MODSS) are specific subcate-
gory within the category of decision support systems. A MODSS contains
formal methods to support Analysis/Evaluation in the selection phase
of a decision process (see figure 1). A wide variety of evaluation
methods such as multicriteria methods, cost-benefit analysis, multi
attribute utility analysis, priority elicitating methods and graphic
evaluation methods are available for analysis and evaluation. Evaluation methods are supplemented with methods for sensitivity analysis to
analyze the sensitivity of evaluation results to various uncertainties
in the available information. Evaluation methods in combination with
methods for sensitivity analysis are used within MODSS to process information available on alternatives and priorities to support evaluation and choice and to support feedback from choice to design and
search. In this way MODSS supports two important cycles in strategic
decision processes:

A. Analysis/Evaluation → Judgement/Bargaining →
  Analysis/Evaluation → etc.

B. Design/Search   → Analysis/Evaluation →
   Judgement/Bargaining → Design/Search   →
   Analysis/Evaluation → etc.

Support of the first cycle results in a better understanding of the
influence of priorities on the final choice and in a better understanding of the reliability of the results. Support of the second cycle
can result in adjustment of existing alternatives, development of new
alternatives and deleting irrelevant alternatives and therefore promo-
tes a complete and adequate set of alternatives. In applications of
MODSS general methods to support Analysis/Evaluation can be linked
with methods specific to the problem to support Design and Search
[5][6].

DEFINITE, the decision support system described in this paper, supports DEcisions with a FINITE set of alternatives in relation to a
finite number of criteria. DEFINITE is commissioned by the policy analysis department of the Dutch Ministry of Finance; intended users are
people involved in the preparation of public planning decisions.
Results range from a complete ranking of alternatives to a graphic
presentation of all relevant information. DEFINITE is to be used in
all sectors of public planning, ranging from stop-go decisions on
infrastructure projects to the selection from various policy schemes
to abate vandalism.

The next section contains the structure and objectives of the DEFINITE
system. Various procedures of DEFINITE are described in section 3
through some examples. Section 4 offers some concluding remarks.

2. DEFINITE STRUCTURE

DEFINITE comprises five modules (see figure 2), each generating their own results. Examples of procedures within each module are given in the next chapter.

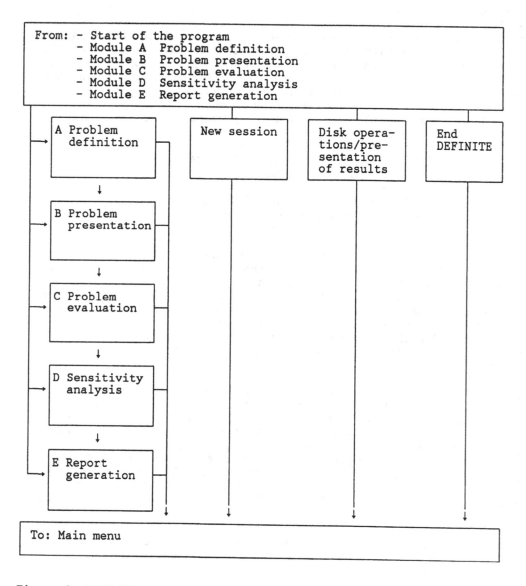

Figure 2. MAIN MENU

Module A, **problem definition**, results in an effects table which represents the decision problem at hand. Module B, **problem presentation**, results in a more concise representation of the effects table. The effects table can be represented as an appraisal table, a graph, a scatter diagram or a cost-benefit sheet.

Module C, **problem evaluation**, generates a ranking of the alternatives. The presence of multiple dimensions, sometimes even of different measurement scales, is handled through multicriteria analysis or cost-benefit analysis. In cost-benefit analysis all scores are converted to a common monetary unit; in multicriteria analysis various forms of standardization are applied which usually results in a common dimensionless unit.

The sensitivities of the rankings obtained in module C to uncertainties in scores, weights and prices, together with the sensitivity of the ranking to the evaluation method used, are analyzed in Module D, **sensitivity analysis**. All results are combined into an evaluation report in Module E, **report generation**. An editor is available to finalize the report.

3. DEFINITE PROCEDURES & EXAMPLES

Each of the five modules contains a number of procedures. All procedures are listed in figure 3, 4, 7 and 9. In this section the most important procedures are described, related to some special issues. A description of all procedures can be found in the User manual [7].

3.1. Module A: Problem definition

The purpose of module A is to obtain an effects table which represents the decision problem at hand. All procedures of Module A are shown in figure 3.

completeness
An important difficulty in problem definition is to reach a complete effects table. Completeness of the set of alternatives may be achieved using procedure A3. The user is requested to identify a number of aspects of the problem at hand and to specify per aspect options for the solution of the problem. The user specifies further which combinations of options are not feasable. The combination procedure now generates all possible alternatives.

example
Problem: Two cities are situated on opposite sides of an estuary. The type of connection between both cities has to be selected.

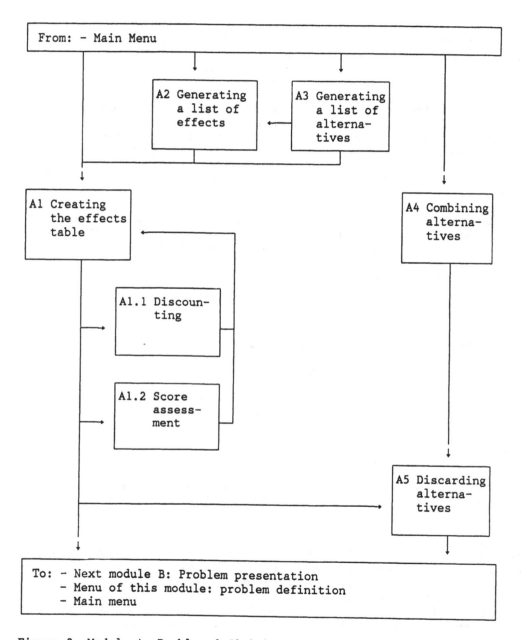

Figure 3. Module A: Problem definition

Table 1 shows three dimensions of this problem and two or three possible solutions for each dimension.

Table 1. Input to the combination procedure

Aspects	Option A	Option B	Option C
Means of transport	Train	Car	
Type of connection	Ferry-boat	Tunnel	Bridge
Capacity (persons/day)	10.000	100.000	

The user specifies that the combination train and ferry-boat and the combination of ferry-boat and a capacity of 100.000 persons/ day are not feasible. The procedure generates ten alternatives (table 2).

Table 2. Output of the combination procedure

Alt 1	Alt 2	Alt 3	''''	Alt 10
Train	Train	Train		Car
Tunnel	Tunnel	Bridge		Bridge
10.000	100.000	10.000		100.000

Completeness of the set of alternatives can also be achieved by aggregating all possible combinations of elements into alternatives. This is performed by procedure (A4). The user specifies elements and combination rules. The procedure generates all combinations of elements that comply with these combination rules.

low information level
Effect scores are not always available as hard numbers. To be able to include all information DEFINITE accepts low information scales such as the ordinal, binary and nominal scale. In table 3 two criteria are scored on a qualitative scale.

example
Problem: Traffic congestions occurs on a road between two cities. There are three possible alternatives to solve this problem. The existing road can be improved, a new road with two lanes route can be constructed (Two lane) or a new highway can be constructed.

Table 3. An effects table

Effects table	Units	Alternatives		
		Improvement	Two lane	Highway
Costs	mln guilders	40	60	80
Travel time	++/--	++	+	+++
Landscape	ordinal	3	1	2
Decrease in accidents	injured/yr	4	5	10

3.2. Module B: Problem Presentation

The purpose of Module B is to present the effects table in a form that allows for the ranking of the alternatives without applying a formal decision rule. All procedures of module B are shown in figure 4.

presentation versus aggregation
Presentation can replace aggregation in problems with a small number of alternatives or criteria. In these cases the overall impression provided by the presentation replaces the scores calculated through decision rules [8].

example
A graphic presentation of table 3 is shown in figure 5. For each criterion the highest bar indicates the best alternative. After ordening the criteria in descending priority the alternatives can be ranked visually (figure 6). DEFINITE can also sort the effects table. In that case a combination of the expected value and the weighted summation method is used.

If an effects table is to large, effects can be combined into effect categories; for example the number of accidents in table 3 may be a weighted sum of the number of accidents with material damage, the number of accidents with injuries and the number of fatal accidents (procedure B2 and B3).

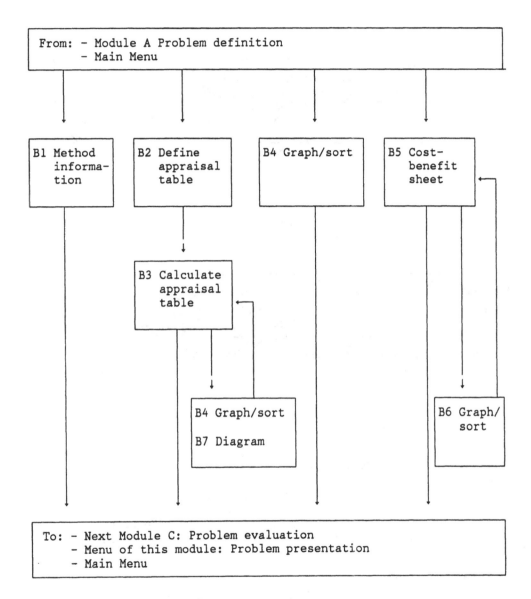

Figure 4. Module B: Problem presentation

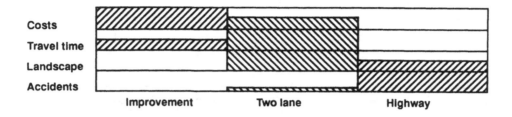

Figure 5. Graphic presentation of an effects table

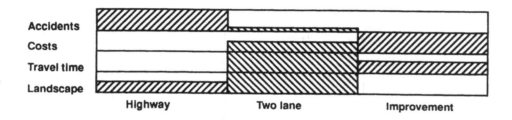

Figure 6. A sorted effects table

3.3. Module C: Problem Evaluation

This module contains two main categories of procedures: monetary evaluation methods [9] and multicriteria methods. All methods result in a ranking of the alternatives by aggregating the effects table according to the decision rules assumed. All procedures of module C are shown in figure 7.

low information level
Multicriteria analysis requires information on priorities. A decision maker is usually only able to express opinions on priorities in a qualitative form. DEFINITE contains various procedures to transform this low level information to quantitative weights as required by the various multicriteria methods.

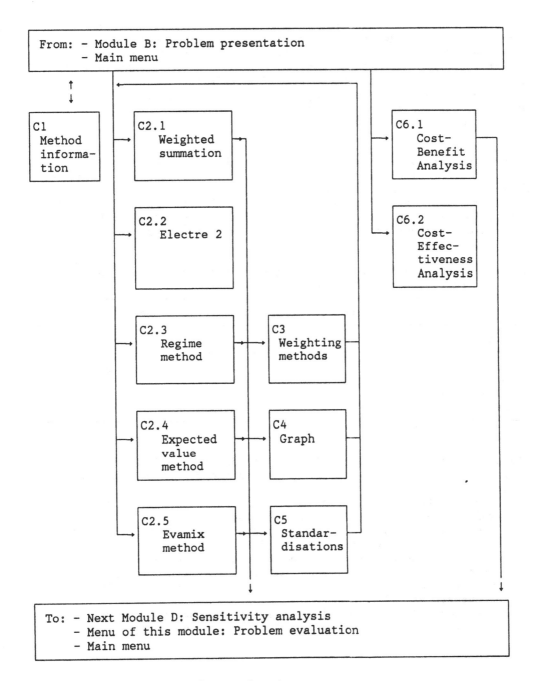

Figure 7. Module C: Problem evaluation

The expected value method, for example, asks the user to order the effects in descending priority order. Each weight is calculated as the expected value of all values conforming to the rank order of that weight [10]. Another method, the Pairwise comparison method, asks the user to compare each pair of effects. The weights are calculated as the Eigenvalues of the resulting matrix of pairwise comparisons [11]. Other methods to deal with qualitative information or priorities are the extreme weights method and the random weights methods [12][13].

example

In the expected value method the user is asked to order the effects in order of decreasing priority e.g. costs are more important than accidents; accidents are more important than travel time etc. In the Pairwise comparison method these questions are asked for all pairs of effects. The user is also asked wether one effect is more important, much more important etc. than the other. The results of both methods are shown in table 4.

Table 4. Weights according to the Expected Value Method and the Pairwise comparison method.

	Expected Value Method	Eigenvalue Method
Ranking	Score	Score
1. Costs	0.521	0.540
2. Decrease in accidents	0.271	0.312
3. Travel time	0.146	0.099
4. Landscape	0.063	0.048

DEFINITE contains five multicriteria methods to transform the effects table in combination with the weights to a ranking of the alternatives:

- The weighted summation method derives the ranking from the weighted sum of standardised effect scores [14].
- The Electre method is based on graphs derived from pairwise comparisons of all alternatives [15][16].
- The regime method is specially designed to handle qualitative or partly qualitative impact matrices. The method is based on partitioning the set of values in accordance to the ordinal effects scores [10].
- The expected value method derives the ranking from the weighted sum of the expected value of the effect scores [17].

- The evamix method ranks the alternatives according to a combination of a dominance index calculated from the qualitative scores and a dominance index calculated from the quantitative scores [13].

The results of these methods, using the weights derived with the expected value method, are presented in figure 8.

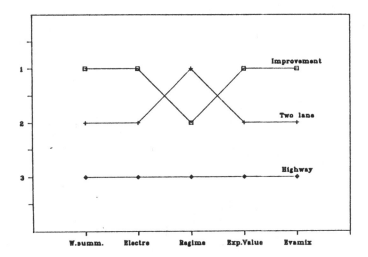

Figure 8. Results of five multicriteria methods

3.4. Module D: Sensitivity analysis

The sensitivities of the rankings obtained in module C to changes in evaluation methods, effects scores and weights are analysed in module D (see figure 9). The module contains procedures to:
- assess the sensitivity of the ranking to the evaluation method applied (method undertainty),
- assess the influence of uncertainties in scores and weights on the ranking of the alternatives (effects and weights uncertainty), and
- to determine the intervals within which the rank order of two alternatives is insensitive to changes in scores or weights (score and weight intervals).
Special search procedures are developed to be able to determine these intervals for all evaluation methods included in the system [18].

negotiations
Score, weight and price intervals can be useful in negotiations where participants hold different views on certain scores or priorities.

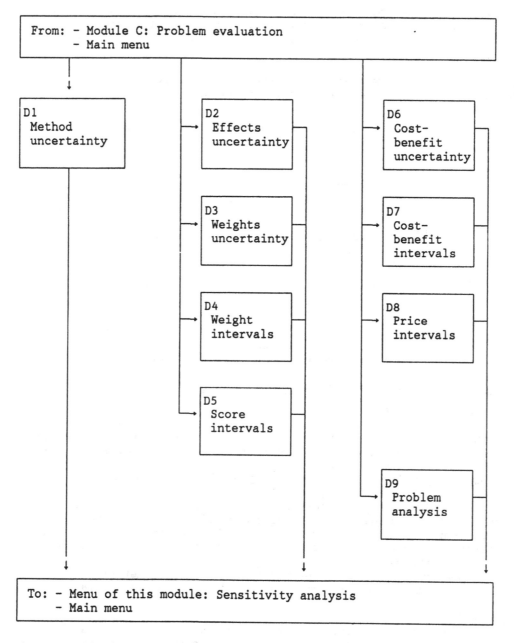

Figure 9. Module D: Sensitivity analysis

example
As shown in figure 8 weighted summation ranks alternative Improvement
before Two Lane and Highway. In figure 10 stability intervals are calculated for these pairs with respect to the number of accidents of
alternative improvement. Figure 10 shows that Two Lane only ranks before Improvement if the number of accidents of Improvement is less
than 0.76 and that, as was to be expected, Highway ranks after Improvement for all possible values of this score.

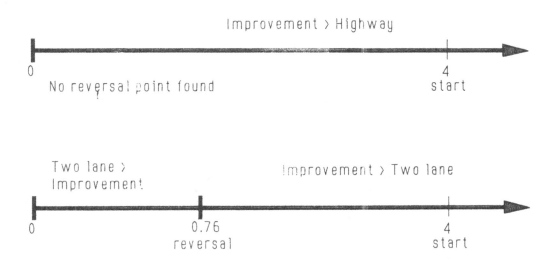

Figure 10. Two score intervals

The same can be done in procedure D4 where DEFINITE calculates stability intervals for weights. The user specifies the weight to be tested
and again two alternatives to compare. The result of this procedure is
presented in figure 11. This figure shows that the ranking of Improvement and Two Lane remains the same until the weight assigned to accidents reaches 0.503 and reverses beyond this point.

Only the relative weight of accidents was changed in calculating the
weight intervals in figure 11; the ratios of all other weights was
held constant. If all weights are allowed to change freely it becomes
clear how sensitive the ranking is to overall changes of the weights.

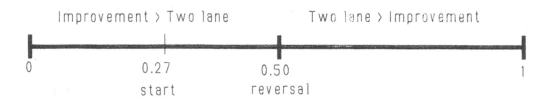

Figure 11. A weight interval

Table 5 shows the weight set with the smallest euclidean distance from
the original weight set that causes a rank reversal of alternatives
Improvement and Two Lane. Table 5 shows that a relative small change
(distance = 0.073) already results in a reversal of the order of the
two alternatives (see also [18]).

Table 5. A weight combination with rank reversal; weighted summation

Original ranking			Reversed ranking (distance = 0.073)		
Weights		Ranking	Weights		Ranking
Costs	0.521	1. Improvement			1. Two lane
Accidents	0.271	2. Two lane			2. Improvement
Travel time	0.146	3. Highway	Travel time		3. Highway
Landscape	0.063		Landscape		

4. CONCLUSIONS

DEFINITE has a bias towards an economic, normative way of decision making. With the exception ofgraphic presentation all rankings
are established through some sort of rational decision rule.

DEFINITE will not be used by decision makers seeking a system that supports other, perhaps more subjective decision procedures. More emphasis should be put on the behavioural aspects of decision making to accomodate for these users [19][20][21].

The use of DEFINITE in public planning has clear advantages. The most important are:
- It allows for support of decision making as a cyclical process [22][23].
- It enables the decision maker to incorporate all relevant information in the decision. It counteracts the unsatisfactory practice of basing a decision only on the number of issues that can be handled by a single individual.
- The flexibility of the system, and specifically the feedback from results to problem content, stimulates "learning by doing". This will increase the understanding of the problem and stimulate the creativity of the decision maker.
- DEFINITE can be used to analyse the influence of changes in data or assumptions on the relative performance of alternatives. This is specifically usefull in group decision making under conflict, where there is usually no agreement on problem definition. The use of DEFINITE can delimit the debate to crucial issues (see also [24]).

There is no doubt that, through an increase in the number of DSS applications, the use of DSS in public planning will increase. This increase will improve flexibility and efficiency of public decision making.

5. LITERATURE

1. Gorry, A. and M.S. Scott Morton: A framework for information systems, Sloan Management Review, vol. 13 (1971), 55-70.
2. Ginzberg, M.J. and E.A. Stohr: Decision Support Systems, in: Issues and Perspective. Ginzberg (eds. M.J. et al), Decision Support Systems, North Holland, Amsterdam 1982.
3. McLean, E. and H.G. Sol (ed.): Decision Support Systems: A decade in perspective, North Holland, Amsterdam 1986.
4. Mintzberg, H., D. Raisinghani, A. Théorêt: The structure of unstructured decision processes, Administrative science quarterly, 21 (1976), 246-275.
5. Janssen, R. and P. Rietveld: Multicriteria Evaluation of landreallotment plans: A case study. Environment and Planning A, vol. 17 (1985), 1653-1668.

6. Janssen, R.: A support system for environmental decisions, in: Evaluation methods for Urban and Regional Plans; Essays in Memory of Morris Hill, Shefer, D. and H. Voogd, Pion, London 1990.

7. Herwijnen, M. van, R. Janssen: DEFINITE. A system to support decision on a finite set of alternatives; User manual, Institute for Environmental Studies, Free University, Amsterdam 1989.

8. Bertin, J.: Graphics and graphic information processing, Walter the Gruyter, Berlin 1981.

9. Mishan, E.J.: Cost-Benefit Analysis, George Allen and Urwin, London 1988.

10. Nijkamp, P., P. Rietveld and H. Voogd: Multicriteria evaluation in physical planning, North Holland, Amsterdam 1990.

11. Saaty, T.L.: The analytical hierarchy process, McGraw Hill, New York 1980.

12. Paelinck, J.H.P.: Qualitative multiple criteria analysis, environmental protection and multi-regional development, Papers of the Regional Science Association, vol. 36 (1974), 59-74.

13. Voogd, H.: Multicriteria Evaluation for urban and regional planning, Pion, London 1983.

14. Keeney, R.L. and Raiffa: Decisions with multiple objectives; preferences and value trade offs, Wiley, New York 1976.

15. Roy, B.: Méthodologie Multicritère d'Aide à la Décision, Economica, Paris 1985.

16. Crama, Y. and P. Hansen: An introduction to the Electre research programme, in: Essays and Surveys on multiple criteria decision making (Ed. P. Hansen), Springer, Berlin 1983.

17. Rietveld, P.: The use of qualitative information in Macro Economic Policy analysis, in: Macro Economic Planning with conflicting goals (Eds. M. Despontin et al.), Springer, Berlin 1984.

18. Rietveld, P. and R. Janssen: Sensitivity analysis in discrete multiple criteria decision problems; on the siting of nuclear power plants. European Journal of Operational Research (forthcoming).

19. Landry, M., D. Pascot, D. Briolat: DSS and the concept of a problem, Decision Support Systems (1985), 25-36.

20. Timmermans, D., C. Vlek: An experimental study of the effectiveness of computer programmed decision support, in: Improving decision making in organisations (Eds. A.C. Locket, G. Islei), Springer, Berlin 1988.

21. Yu, P.L.: Understanding behaviors and Forming Winning Strategies. An integrated theory of habitual domains, University of Kansas, Kansas 1990.

22. Faludi, A.: Planning theory, Pergamon Press, Oxford 1971.

23. Simon, H.A.: Administrative Behaviour: A study of Decision making Processes in Administrative Organization, New York 1970.

24. Janssen, R. and W. Hafkamp: A decision Support system for Conflict Analysis. The annals of regional science, vol. 20 (1986), no. 3, 67-85.

TWO DECISION SUPPORT SYSTEMS FOR CONTINUOUS AND DISCRETE MULTIPLE CRITERIA DECISION MAKING: VIG AND VIMDA

P. Korhonen

Helsinki School of Economics, Helsinki, Finland

The research is supported, in part, by grants from the Foundation of the Helsinki School of Economics, and the Foundation of the Student Union of the Helsinki School of Economics.

ABSTRACT

Two "free search" approaches and their implementations (VIG and VIMDA) for solving continuous and discrete multiple criteria problems are considered. The systems are user-friendly and based on visual interaction. The user communicates with the respective systems using spreadsheets, menus, and graphics. The mathematical foundations are based on the reference direction approaches. The basic underlying ideas of the approaches are described and the use of the systems is demonstrated. Several real applications of VIG will be discussed, for example, how to analyze input-output models from the perspective of economic defense.

Keywords: Multiple Criteria, Decision Support, Visual, Interactive, Computer Graphics, Reference Direction

1. INTRODUCTION

As Weber [1986] points out, the concept of Decision Support Systems (DSS) is now more than 15 years old. Decision Support Systems have over the years been developed and successfully implemented in many organizations by individuals at different organizational levels. Various definitions have been offered for decision support. Ginzberg and Stohr [1982] define a DSS as "a computer-based information system used to support decision-making activities in situations where it is not possible or not desirable to have an automated system perform the entire decision process".

Multiple Criteria Decision Support Systems (MCDSS) are considered by Jelassi et al. [1985] as a "specific" type of system within the broad family of DSS. Even though MCDSS include much the same components as "traditional" Decision Support Systems, MCDSS have special characteristics that distinguish them from other DSS. Such characteristics include the fact that they allow analysis of multiple criteria; they use a variety of multiple criteria decision methods to compute efficient solutions; they incorporate user's input in various phases of modelling and solving a problem. While it has been customary to consider algorithms as the focal point of decision support, emphasis is shifting to the database and modelling activities including the choice behavior of a decision maker (DM). See, for example, Keen and Scott-Morton [1978], Sprague and Carlson [1982], Bui [1984], and Van Hee [1987].

To solve multiple criteria decision problems requires the intervention of the DM. Therefore, the most MCDSS are interactive. Several dozen procedures have been developed for solving both continuous and discrete optimization problems having multiple criteria. For an excellent review of several interactive multiple criteria procedures, see Steuer [1986].

The specifics of these procedures vary, but they have several common characteristics. For example, at each iteration, a solution, or set of solutions, is generated for a decision-maker's (DM's) examination. As a result of the examination, the DM inputs information in the form of tradeoffs, pairwise comparisons, aspiration levels, etc. His/her responses are used to generate a presumably, improved solution. The ultimate goal is to find the so-called most preferred solution of the DM. Which search technique and termination rule is used is heavily dependent on the assumptions about the behavior of the DM and the way in which these assumptions are implemented. In the MCDM-research there is a growing interest in the behavioral realism of such assumptions.

Currently, none of the MCDSS fully meet the requirements of behavioral realism. Even in a "free search" type of systems   like VIG (A Visual Interactive Goal Programming) (Korhonen, 1987a) and VIMDA (A Visual Interactive Method for Discrete Alternatives) (Korhonen, 1988), the sequence of the solutions to be considered obviously effects the choice of the final solution. However, this principle does not restrict the choices of the DM, and provides him/her with a possibility to consider any nondominated solution he/she pleases. The DM can feel that the system is totally under his/her control making him/her satisfied with the final solution. (If this solution is really the "best" solution to the DM, it is "philosophical question", to which we cannot answer at this moment.)

The foundations of VIG originate in the visual interactive reference direction approach to multiple objective linear programming developed by Korhonen and Laakso [1986a and 1986b]. Korhonen [1988] extended the original idea into a discrete alternative problem and implemented it by name VIMDA. The main idea in the reference direction approach is to project a given direction onto the set of the efficient frontier, and thus find an ordered subset of efficient solutions (infinite or finite) for the DM's evaluation. The whole subset, or parts thereof, is presented to the DM using computer graphics.

Moreover, Korhonen and Wallenius [1988] developed a method to specify a reference direction in a dynamic way. This method is implemented in VIG by name PARETO RACE. In PARETO RACE the DM can move in any direction (on the efficient frontier) he/she likes and no unduly restrictive assumptions concerning the DM's behavior are made. Various directions are specified by using function keys making the DM feel that he/she is like "driving a car" on the efficient frontier.

VIMDA uses the original interface proposed by Korhonen and Laakso [1986a] to present on a screen the whole subset of alternatives corresponding to a certain reference direction. By moving the cursor back and forth, the DM is encouraged to consider alternatives belonging to this subset and make his/her choice. The chosen alternative becomes the basis for the subsequent iteration. The DM will continue in this manner until he/she has found an alternative that satisfies him/her.

The both of the systems are user-friendly. Menus, spreadsheets, and interactive use of computer graphics play a central role. The key concept is visual interaction. The systems are written in TURBO PASCAL and implemented on an IBM PC/1 microcomputer. To benefit fully from the systems, a color monitor should be used. In addition, VIMDA requires a graphics card.

This paper consists of seven sections. The introduction explains the underlying philosophy of the systems (VIG and VIMDA) and a brief history. The second section provides the mathematical foundations. The third section describes the structure of the systems. The fourth section discusses PARETO RACE, which is an essential part of VIG implementing visual interaction. The fifth section describes the computer-graphics interface in VIMDA. In the sixth section several applications of VIG are described. The seventh section concludes the paper.

2. MATHEMATICAL FOUNDATIONS

2.1. A Generalized Goal Programming Model

Let us consider the decision problem, where the consequences (outcomes) y_i, i = 1,..., m, of decisions (activities, actions, choices) are described as functions of the decision variables x_j, j=1,...,n:

$$y = f(x), \tag{2.1}$$

in which function $f(x) = (f_k(x))$: $R^n \rightarrow R^m$, k ϵ I, is a mapping from the decision variable space into the space of consequences. Vector x is called a decision vector and y a consequence vector.

Using multiple objective mathematical programming for solving model (2.1), some of the outcome variables are chosen as objectives which are to be maximized or minimized within some constraints. The constraints are defined by specifying acceptable values for the other outcome variables. In the subsequent discussion, we assume that the vector y includes the vector x, when the DM imposes restrictions upon it or has preferences over the values of the decision variables.

Often, the objectives are more convenient to be considered as goals by setting desirable values (aspiration levels) for objectives which are not absolute as for constraints. We may assume that the DM wishes to improve (in our formulation to maximize) the values of the objective functions corresponding to such goals as much as possible.

Following Ignizio [1983], constraints may be regarded as a subset of goals: a constraint is an **inflexible** goal. Accordingly, constraints and goals may be formulated in a uniform manner (see, Korhonen and Laakso [1986b]):

$$y_i = f_i(x) \geq b_i, \; i \; \epsilon \; G,$$
$$y_j = f_j(x) \geq b_j, \; j \; \epsilon \; R, \tag{2.2}$$

where G is the index set of flexible goals, and R is the index set of inflexible goals (constraints). The index set of all goals (flexible and inflexible) is denoted by I, I = G$\cup$R. Vector b_k, k ϵ I, is an aspiration level corresponding to each consequence. For each goal (flexible or inflexible), we can define a deviation variable $d_k = b_k - f_k(x)$. If $d_k \leq 0$, then the kth goal is inflexible. If d_k is unbounded, the kth goal is flexible. In the case of inflexible goals, values exceeding aspiration levels have no special meaning.

The generalized goal programming problem can be formulated as follows:

$$\min d_i \text{ for all } i \in G,$$

subject to: (2.3)

$$f_i(x) + d_i = b_i, i \in I,$$
$$d_j \leq 0, \text{ for } j \in R.$$

The inflexible goals define the feasible set X_R:

$$X_R = \{x \mid f_j(x) + d_j = b_j, d_j \leq 0 , j \in R\},$$

and an $x \in X_R$ is called a feasible point.

We define efficiency and weak efficiency in the usual manner:

<u>Definition 1.</u> A point $x° \in X_R$ is efficient iff there does not exist another $x \in X_R$ such that $f_i(x) \geq f_i(x°)$ for all $i \in G$ and $f_i(x) \neq f_i(x°)$ for at least one $i \in G$.

<u>Definition 2.</u> A point $x° \in X_R$ is weakly-efficient iff there does not exist another $x \in X_R$ such that $f_i(x) > f_i(x°)$ for all $i \in G$.

The criterion vectors corresponding to (weakly) efficient points are called (weakly) efficient solutions.

2.2. A Reference Point and Direction

In order to generate efficient solutions for the DM's evaluation, some rules bearing a relationship to the DM's aspirations must be established. Such rules may be based on an implicitly or explicitly defined **achievement (scalarizing) function** as suggested by Wierzbicki [1980]. An efficient solution can be found by optimizing some achievement function of the goal deviation variables d_j, $j \in G$. An achievement function may be specified in several ways. We use the following simple function:

$$z(d,w,G) = \max \{d_i / w_i, i \in G\},$$ (2.4)

where w is a positive weighting vector. By minimizing $z(d,w,G)$ we find a weakly-efficient point (see, e.g., Wierzbicki [1980] and [1986]). The simple form is adequate for our purposes, because, technically, all we need is a handy "efficient solution generator". Function (2.4) performs precisely this operation.

Given a feasible point x^* that is known to the DM, we can define a **reference direction** as any direction starting at $f_i(x^*)$, $i \in G$, and offering a preferable change in the values of the objective functions (see, Korhonen and Laakso [1986a]). One way to specify a reference direction is simply to ask the DM to give aspiration levels g_i, $i \in G$, for flexible goals and assume that total utility increases in the direction $v_i = g_i - f_i(x^*)$.

By using the achievement function (2.4), we can project a reference direction on the set of weakly-efficient solutions by solving the parametric programming problem:

$$\min\ \max\ \{[(f_i(x^*) + t(g_i - f_i(x^*)) - f_i(x)]/w_i\ ,\ i\ \varepsilon\ G\}$$

subject to: $\hspace{9cm}$ (2.5)

$$f_j(x) \geq b_j\ ,\ j\ \varepsilon\ R,$$

where $t \geq 0$. Problem (2.5) can be converted into the following equivalent form:

$$\min\ \varepsilon$$

subject to: $\hspace{9cm}$ (2.6)

$$
\begin{aligned}
f_i(x) +\ \ w_i\ \varepsilon\ &\geq\ f_i(x^*) + t(g_i - f_i(x^*))\ ,\ \text{for all}\ i\ \varepsilon\ G, \\
f_j(x)\ \ \ &\geq\ b_j\ ,\ j\ \varepsilon\ R, \\
&\varepsilon\ \text{is unrestricted,}
\end{aligned}
$$

where ε is a scalar variable to be minimized and $t \geq 0$ is a parameter. When $t = 1$, the model finds a feasible solution for the given aspiration levels g_i, $i\ \varepsilon\ G$.

It is easy to see that model (2.6) can be also given in a general form:

$$\min\ \varepsilon$$

subject to: $\hspace{9cm}$ (2.7)

$$
\begin{aligned}
f_i(x)\ +\ w_i\ \varepsilon\ &\geq\ b_i + tr_i\ ,\ \text{for all}\ i\ \varepsilon\ G, \\
f_j(x)\ \ \ &\geq\ b_j\ ,\ j\ \varepsilon\ R, \\
&\varepsilon\ \text{is unrestricted.}
\end{aligned}
$$

In model (2.7), the starting solution $f_i(x^*)$ for search is found by projecting the aspiration levels b_i, $i\ \varepsilon\ G$, on the efficient frontier: $f_i(x^*) = b_i - w_i\varepsilon^* - z_i$, where ε^* gives an optimum for the model (2.7) with $t = 0$, and z_i is a surplus variables corresponding to row i. The reference direction vector $r = (r_i)$, $i\ \varepsilon\ G$, is any direction vector specified by the DM.

By dynamically varying the reference direction, the DM is able to examine any part of the efficient frontier.

2.3. A Linear Decision Model

Let us now consider the problem, where the consequences y_i, $i = 1,..., m$, of decisions can be stated as linear functions of the decision variables x_j, $j=1,...,n$:

$$y_i = y_i(x) = \sum_{j=1}^{n} a_{ij}x_j\ ,\ i\ \varepsilon\ M = \{1,2,...,m\},$$

or equivalently in the matrix form:

$$y\ = y(x) = Ax,$$

where A is an m x n matrix of coefficients. The problem is to find values for the decision variables x_j, $j \in N = \{1,2,...,n\}$, such that the outcome variables, y_i, $i \in M$, would have acceptable or desirable values. If $n \geq m$, then for each desired or given value of y, there exist an infinite number of solutions for the model, and it can easily be solved. To avoid this trivial case, we assume that $m > n$.

Corresponding to formulation (2.7), the linear model can be given as

$$\min \varepsilon$$

s.t.   (2.8)

$$\begin{aligned} A_1 x + w\varepsilon - Iz_1 \quad &= b_1 + tr \\ A_2 x \quad\quad\quad - Iz_2 &= b_2 \\ z_1, \quad z_2 &\geq 0, \end{aligned}$$

where $w > 0$ is a k-vector of weights, $r \in R^k$ is a reference direction, $z_1 \in R^k$ and $z_2 \in R^{m-k}$ are the vectors of surplus variables for flexible goals and inflexible goals, b_1 and b_2 are corresponding aspiration level vectors, and $t \geq 0$ a parameter.

Furthermore, it is not necessary to make a conceptual distinction between objective functions and constraints, because their roles can be changed during the solution process. A unified formulation for objective functions and constraints can be written as follows (see, Korhonen and Wallenius [1988]):

$$\min \varepsilon$$

subject to: (2.9)

$$\begin{aligned} Ax + w\varepsilon - Iz &= b + tr, \\ z &\geq 0, \end{aligned}$$

where $z \in R^m$ is the vector of surplus variables for all rows, $w \geq 0$, $w \in R^m$, is now a vector whose components are

$$w_i \quad \begin{bmatrix} = 0, \text{ if i refers to a constraint,} \\ > 0, \text{ if i refers to an objective,} \end{bmatrix}$$

and $r \in R^m$ is a reference vector with

$$r_i \quad \begin{bmatrix} = 0, \text{ if i refers to a constraint,} \\ \neq 0, \text{ if i refers to an objective,} \end{bmatrix}$$

and $b \in R^m$ is a vector consisting of aspiration levels for flexible and inflexible goals.

After finding an initial solution for model (2.9) with $t = 0$, we can use linear parametric programming to move on the efficient frontier. For each reference direction r and t: $0 \rightarrow \infty$, we will find an efficient curve consisting of an infinite number of efficient solution. By varying a reference direction and controlling the value of parameter t, we can implement different versions for presenting these solutions for the DM's evaluation.

In the VIG system, the DM can dynamically change a reference direction and control parameter t as explained in section 4.

2.4. A Model for Discrete Alternatives

In a discrete multiple criteria decision problem, we assume that there is a single DM, a set of n deterministic decision alternatives, and m criteria $(m > 1)$, which define an $n \times m$ decision matrix A whose elements are denoted by a_{ij}, $i \in I = \{1,2,...,n\}$ and $j \in J = \{1,2,...,m\}$. We use a_i or i to refer to the decision alternative in row i. Thus, each decision alternative is a point in the criterion space R^m.

Assuming that a DM wishes to maximize each of the m criteria, the problem is to

$$"max"\ a_i. \qquad (2.10)$$
$$i \in I$$

A discrete problem can be presented as a special case of the general model (2.9) by adding the following extra constraints:

$$min\ \varepsilon$$

subject to: $\qquad\qquad\qquad\qquad\qquad\qquad\qquad\qquad (2.11)$

$$A'x + w\varepsilon - \quad Iz = b + tr,$$
$$z \geq 0,$$
$$1'x \qquad = 1$$
$$x_i \qquad = 0\ or\ 1,\quad for\ i = 1,2,\ ...,\ n,$$

where $t \geq 0$, A' stands for the transpose of matrix A and the vector with each component equal to one is denoted by 1.

Formulation (2.11) specifies that any of the nondominated columns (rows) of matrix A' (A) is a possible solution for the problem. In the discrete problem, the dominated alternatives are easy to eliminate before starting a solution process. By solving the parametric programming problem (2.11), t: $0 \rightarrow \infty$, we obtain an ordered set of decision alternatives a_i, $i \in K \subseteq N$, as a solution. This set is presented for the DM's evaluation. The problem is not easy to solve. An algorithm is described in Korhonen [1988].

In the VIMDA system, the specification of the reference direction is implemented according to the original idea of Korhonen and Laakso [1986a]. The aspiration levels of the criteria are used for this purpose. The DM will specify the aspiration levels for criteria and the reference direction is specified as a direction starting from the current solution and passing through the aspiration levels.

3. THE DESIGN PRINCIPLES AND STRUCTURE OF THE SYSTEMS

The main aim in designing systems VIG and VIMDA has been to develop two user-friendly multiple criteria decision support systems which help the DM analyze efficient solutions without mathematical or computer "jargon".

The DM needs to understand the concept of efficiency, but he/she does not need to know how the system uses mathematics necessary for generating efficient solutions. The logic of the system is transparent. The DM can see what the system does, but he/she does not know how. It is also important that the DM can control the system making him/her feel that he/she is not at the mercy of the system.

The key concept is "free search". The DM can freely to control the search of efficient solutions. The systems allow him/her to perform a "free search", without unnecessary mathematical restrictions.

To fulfill these requirements the systems are implemented in the spirit of visual interaction. "Visual interaction" means that the DM communicates with the system using visual representation in an interactive manner. Holistic as well as detailed information is presented, simultaneously. Of course, computer graphics play a central role in visual interaction.

Furthermore, the systems are compatible with recent findings in the psychology of decision-making. Such findings indicate that DMs are not always consistent or transitive and that they may change their mind about the quality of solutions.

Both of the systems consist of more than 3500 lines of code. Most of it is used to build up an attractive user/computer interface. The interface is based on one main menu, spreadsheets, computer graphics, and the use of colors.

The current version of VIG is capable of solving problems with a maximum of 95 columns and 100 rows, from which at most 10 may be defined as objectives, simultaneously. You can change the role of objectives and constraints during a session.

Using VIMDA you can solve problems with a maximum of 10 attributes (criteria) and 400 alternatives.

Logically, the structure of the systems consists of four main functions:

Data (Model) Management Function

> which allows the DM to create new data, retrieve existing data, and save modified and restructured data. Furthermore, he/she may store the currently best solution at each iteration.

Data (Model) Operations Function

> which allows the DM to provide the attributes (columns) and the alternatives (rows) with names, to create new alternatives and attributes, to fill in or edit the attribute-alternative matrix, to specify acceptable ranges for the attributes, and to perform various preliminary operations with his/her data.

Problem Solving Function

which allows the DM to specify, how he/she want his/her problem to be solved, and then actually solve it. It consists of defining the decision-relevant criteria, ranges for the criteria, and the procedure (PARETO RACE in VIG and Computer-Graphics approach in VIMDA) itself for solving the problem. If the DM has specified only one criterion, no what-so-ever intervention of the user is needed to solve the problem. Otherwise, the graphical reference direction approach is evoked to support him/her to find his/her most preferred solution.

Solution Output Function

which allows the DM to examine solutions on the screen, and store intermediate results for later consideration and/or for report writing. The file is a standard ASCII-file and thus suitable to any textprocessing system.

Each main function consists of one or more sub-functions, which all are selected from the main menu. At each stage, available choices are shown using a special color (= light cyan). The interface of the main menu of the systems VIG and VIMDA are quite similar. Figure 1 will illustrate the main menu of VIG.

```
                         V I G
                         ↑↑↑↑↑↑↑

          The Name of the Model:     <None>

                         Menu
                         ↑↑↑↑

          ==>  Select Model
               Name Rows and Columns
               Edit Matrix Coefficients
               Specify the Types and Values of Goals
               Save the Model
               Classify the Goals
               Specify Ranges for Flexible Goals
               Solve the Problem
               Display the Values of Goals
               Display the Values of Decision Variables
               Save the Current Solution
               END

< ⌐:Proceed F10:As < ⌐
```

Figure 1. The Main Menu of VIG

The details of the use of the systems are described in VIG's and VIMDA's User's Guides (see, Korhonen, [1987b] and Korhonen and Wallenius, [1989]).

4. PARETO RACE

PARETO RACE is an essential part of system VIG implementing visual interaction. In PARETO RACE, you can freely search the efficient frontier of a multiple objective linear programming problem. Specific keys are used to control the speed and direction of motion. On a display, you see the objective function values in numeric form and as bar graphs whose lengths are dynamically changing as you move about on the efficient frontier. The keyboard controls include the following function keys (see, Fig. 2):

(SPACE) BAR: An "Accelerator"

> You proceed in the current direction at constant speed.

F1: "Gears (Backward)"

> Increase speed in the backward direction.

F2: "Gears (Forward)"

> Increase speed in the forward direction.

F3: "Fix the Aspiration Level of a Specific Goal"

> The current value of a specific goal is taken as an absolute lower (upper) bound.

F4: "Relax the Aspiration Level of a Fixed Goal"

> The fixed goal becomes flexible again.

F5: "Brakes"

> Reduce speed.

num: "Turn"

> Change the direction of motion by pressing the number key corresponding to the goal's ordinal number once or several times.

F10: "EXIT"

> Exit to the main menu.

Initially, you are going to see (graphically and numerically) an efficient solution on the computer screen (Fig. 2). Arrows indicate an initial direction chosen by the computer (based on our aspiration levels). If you like this initial direction, you hold SPACE BAR down and observe the solutions change. You are "travelling" at base speed. If you like to increase the speed, you press the F2-key once or several times (depending on the desired speed) and hold SPACE BAR down. If at some point the direction is no longer attractive, you initiate a turn (see, Fig. 2). Assume that the user wishes to improve a certain goal. To accomplish this, the goal's corresponding number key is pressed once or several times, depending on the amount of desired improvement. Then, the program updates the direction, and so forth. To reduce speed, you use brakes (F5-key). If you wish to resume base speed, the most convenient way is to press the F1-key (gears: backward) and then the F2-key (gears: forward). You reverse direction twice and start travelling at base speed. When a flexible goal is fixed by using the F3-key, the system asks us to specify the number of the goal to be fixed; then an asterisk appears next to the name of such a goal. The goal may be relaxed by using the F4-key; the asterisk disappears.

```
                                   Pareto Race



       Goal    1 (min ): Mach.Hours <==
   ██████      8.4286
   ▐Goal       2 (min ): Man Hours  ==>
   ███         9.4286
   ▐Goal       3 (max ): Profit      <==
   ████████████████████      29.1429




Bar:Accelerator F1:Gears (B)  F3:Fix     num:Turn
F5:Brakes       F2:Gears (F)  F4:Relax  F10:Exit  * Goal # 3 is improved *
```

Figure 2. An Example of PARETO RACE: Initial Solution (Making a Turn)

When you terminate PARETO RACE, you are youlcome to examine the values of the decision variables and goals. Just exit to the main menu. In fact, the user can examine the values of the decision variables and goals also during the race (and save such intermediate results for later use). Simply enter "Solve the Problem" in the main menu and you are back in the race. You can also change the role of flexible and rigid goals during the race. You feel that this adds flexibility to the system and extends it beyond "classical" multiple objective linear programming.

5. A COMPUTER-GRAPHICS INTERFACE IN VIMDA

When the DM has specified more than one decision-relevant criterion for his/her discrete decision problem (the usual case), he/she needs the reference direction approach to find the most preferred solution. He/she is guided to enter the right selection from the main menu, and the system will display a screen giving information of the current alternative and the ranges of the criteria, and provides the DM with a possibility to give the aspiration levels for his/her criteria. After this information, the system can build up a reference direction and generate an representative (ordered) subset for the DM's evaluation. The initial screen of the reference direction approach is displayed (see, Fig. 3).

The current alternative is shown on the left hand margin. The criterion values of consecutive alternatives have been connected with lines using different colors and patterns.

The cursor refers to that alternative whose criterion values are printed numerically on top of the screen. The user is asked to choose the most preferred alternative from the screen by moving the cursor to point to such an alternative.

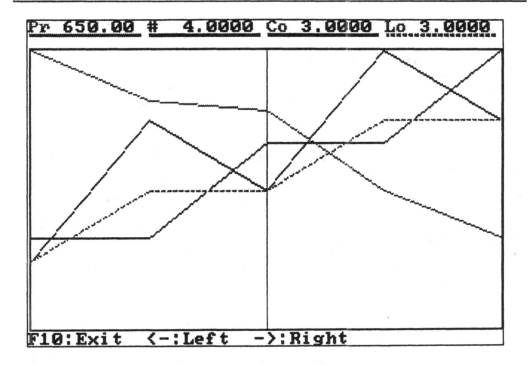

Figure 3. The Screen of the Computer-Graphics Interface in VIMDA

The chosen (most preferred) alternative becomes the current solution for the next iteration. Next, the user is asked to reconsider his/her aspiration levels, etc.

The process stops when the DM is satisfied with the solution.

6. THE APPLICATIONS OF VIG

The production version of VIG has been available for some years, whereas VIMDA was launched for public use last year. Therefore, so far we have only experiences on applying VIG to solving several practical problems. In this section you briefly survey a number of such applications. Additional details may be found in the original references. Up till now, most of the problems solved have been relatively small-scale (with respect to the number of rows and columns), but there clearly exists a need (and the potential) for solving large-scale problems using VIG.

One of the earliest applications deals with the determination of optimal price changes for alcoholic beverages sold by ALKO Ltd., the Finnish State Alcohol Monopoly. Pricing decisions are among the most important ones the company must make. The objective is not only to maximize profits, but to restrict sales with the intention of reducing harmful effects of alcohol consumption, and to minimize the impact of price increases on the consumer price index. A linear programming model with three objective functions (reflecting the above concerns) was developed and solved using our procedure. The decision variables of the model youre logarithms of relative price changes. Modifications in the input-output routines of the program enabled the user to operate with original percentages instead of their logarithms. Based on this model, the company has implemented an extended version of the model. The ultimate DM, the governmental ministers comprising the cabinet, is interested in the model. Additional details may be found in Korhonen and Soismaa [1988].

Another interesting and important application concerns the problem of stockpiling critical materials to be used in the event of a national emergency in Finland. It is essentially a bi-criteria problem. On the one hand, the purpose is to minimize the cost of stockpiling, and on the other hand to maximize the youlfare of the population at large during (and after) a crisis. Using the model it is possible to make intelligent decisions regarding appropriate stockpiling levels for critical materials, and understand the implications in case of a crisis. The National Board of Economic Defense has presented a document on this project to the Ministry of Trade and Industry, and the system is being used by the board on a regular basis. Because of the critical nature of this project, no scientific articles have been published on the system, but the system and its merits have been summarized in some Finnish newspapers.

In Kananen et al. [1990] you demonstrate how VIG can effectively be used for analyzing input-output models. You have applied our approach to studying the effects of economic or political crises on the Finnish economy. Examples of such crises are nuclear poyour plant accidents, trade embargoes, and international conflicts of various nature. An input-output model, based on the latest official input-output statistics of the Finnish economy with 17 industries (sectors) is employed. Relevant objective functions include maximization of (private) consumption, employment, and total gross outlay. Our system has been implemented on a microcomputer and is currently being used on a regular basis by the National Board of Economic Defense. For large-scale problems, such as inter-temporal, regional, input-output models, a large-scale version of VIG should be used.

An interesting first application of VIG is being performed for a major Finnish wholesaler of hardware and related products. The problem concerns a division of a company that has several departments or units. The analysis starts from the income statements of each department. Each of these income statements forms a column in our model. Some of the items (rows) in the income statements (for example, profits, turnover, etc.) are defined as objective functions (flexible goals). Using VIG the company management can decide, what they want to achieve (division-wise) in terms of profits, turnover, and so forth. Then, they can find out, which departments are instrumental (critical) in achieving these overall goals, and which departments need strengthening in terms of managerial skills, capital, and labor. Interestingly, our system identified certain youak departments in the division (confirming the suspicions of management) and identified some strategic opportunities for other departments. Based on this study, the company has reached a decision to start using VIG at management level in the future. The results will be forthcoming as a future report.

You have applied VIG to selecting advertising media for a Finnish software company. The problem was to assist the management in allocating an advertising budget across various media. The purpose was to maximize audience exposure. Six different newspapers and professional magazines youre initially chosen as the relevant advertising media. The relevant audience consisted of the following (target) groups: Marketing Management, Finance and Personnel, ADP, Production Management, R&D, and General Management. To measure the audience exposure, readership numbers youre obtained from Finnish Gallup Ltd., reflecting the current situation. The media selection problem was modelled as a linear decision problem in the spirit of VIG. The decision variables youre the numbers of ads in each medium, which youre also considered as consequences in our model. For additional details, an interested reader may consult Korhonen et al. [1989].

You would also like to mention the work done by Joen Forest Consulting, a Finnish consulting company specialized in forest sector management models. They are using VIG as an important part of their system for forest planning, harvesting, and replanting. The system is being used by more than half a dozen forest sector (vocational) educational institutions in Finland. It has also been used for a forest project in Zambia. (See Pukkala, et al., 1989.)

The management of NumPlan, Ltd., a small Finnish software company that markets the VIG software has also used VIG in planning its marketing strategy. Alternative strategies youre identified and used in a multiple objective linear programming model involving qualitative data. Based on the analysis and use of the VIG program, several strategies appeared which have substantially changed our approach to marketing VIG. See Korhonen et al. [1990].

You also cite several studies that have been made, that have not yet resulted in concrete applications.

The waters of the New York Bight are among the most intensively utilized in the world. Concern for water quality in this region is long standing. Yet, sewage sludge has been dumped at the so-called 12-mile site for more than 60 years. Now the communities using the site are shifting their operations to the more distant 106-mile site, following orders issued in 1985 by the U.S. Environmental Protection Agency. In a recent study, Wallenius et al. [1987] used VIG to re-examine the EPA decision in a way which permits simultaneous multi-site dumping. The study is based on a linear programming model, providing a framework for ocean waste disposal management. The relevant objective functions youre to minimize transportation costs to New York City, and communities in New Jersey and on Long Island, and to minimize the volume of pollutants in near-shore and off-shore areas. The decision variables in the model are the number of barge trips made from source i to site j by disposal method k. The model consists of 40 decision variables and 27 constraints (inflexible goals) imposing environmental and capacity restrictions.

Duckstein et al. [1988] describe the use of VIG in forest management. More specifically, they examine the case of managing a ponderosa pine forest in the Salt-Verde river basin, situated in North-Central Arizona. The problem is to allocate to given subwatersheds six different treatment strategies corresponding to clearcutting betyouen 0 and 100% of the area. This allocation is evaluated using multiple, conflicting objective functions, such as water yield, sediment yield, amount of recreational usage, economic benefits/costs. The problem has a number of physical constraints, some of which are fuzzy. For additional details, see Duckstein et al. [1988].

7. CONCLUSION

In this paper you have described two software packages called VIG and VIMDA, which are developed for supporting multiple criteria decision making. VIG is a system for modelling and solving multiple objective linear programming problems, and VIMDA is meant for the problems, where the alternatives are explicitly enumerated.

Our objective is to make the programs widely available, and pursue several applications.

References

Bui, X. T. (1984), "Building Effective Multiple Criteria Decision Support Systems", Systems, Objectives, Solutions 4, Nr. 1, 3-16.

Duckstein, L., Korhonen, P., and Tecle A. (1988): "Multiobjective Forest Management: A Visual, Interactive, and Fuzzy Approach", in Kent, B. M. and Davis, L. S. (Eds.): Proceedings of the 1988 Symposium on Systems Analysis in Forest Resources, Asilomar Conference Center, Pacific Grove, California, March 29-April 1, 1988.

Ginzberg, M. and Stohr, E. (1982), "Decision Support Systems: Issues and Perspectives", in: M. Ginzberg, W. Reitman, W., and E. Stohr, Eds., Decision Support Systems, (North-Holland Publ. Comp., Amsterdam, 1982).

Ignizio, J. P. (1983), "Generalized Goal Programming", Computers and Operations Research 10, 277-289.

Jelassi, M. T., Jarke, M. and Stohr, E. (1985), "Designing a Generalized Multiple Criteria Decision Support System", Journal of Management Information Systems I, 24-43.

Kananen, I., Korhonen, P., Wallenius, H. and Wallenius, J. (1990): "Multiple Objective Analysis of Input-Output Models for Emergency Management", Operations Research 38, Nr 2, March - April 1990.

Keen, P. G. W. and Scott-Morton, M. (1978), Decision Support Systems: An Organizational Perspective, Addison Yousley.

Korhonen, P. (1987a), "VIG - A Visual Interactive Support System for Multiple Criteria Decision Making", Belgian Journal of Operations Research, Statistics and Computer Science 27, Nr. 1, 3-15.

Korhonen, P. (1987b), VIG (A Visual Interactive Approach to Goal Programming) -User's Guide, NumPlan Ltd.

Korhonen, P. (1988), "A Visual Reference Direction Approach to Solving Discrete Multiple Criteria Problems", European Journal of Operational Research 34, Nr. 2, 152-159.

Korhonen, P., and Laakso, J. (1986a), "A Visual Interactive Method for Solving the Multiple Criteria Problem", European Journal of Operational Research 24, Nr. 2, 277-287.

Korhonen, P. and Laakso, J. (1986b), "Solving Generalized Goal Programming Problems Using a Visual Interactive Approach", European Journal of Operational Research 26, Nr. 3, 355-363.

Korhonen, P., Narula, S. and Wallenius, J. (1989): "An Evolutionary Approach to Decision-Making, with an Application to Media Selection", Mathematical and Computer Modelling 12, Nr. 10/11, 1239-1244.

Korhonen, P., Siljamäki, A. and Wallenius, J. (1990): A Multiple Criteria Decision Support System VIG in Corporate Planning, NumPlan Ltd.

Korhonen, P. and Soismaa, M. (1988): "A Multiple Criteria Model for Pricing Alcoholic Beverages", European Journal of Operational Research 37, Nr. 2, 165-175.

Korhonen, P. and Wallenius, J. (1988), "A Pareto Race", Naval Research Logistics 35, Nr. 6, 615-623.

Korhonen, P. and Wallenius, J. (1989), A Visual Multiple Criteria Decisionb Support System VIMDA For Discrete Alternatives with Numerical Data -User's Guide, NumPlan Ltd.

Pukkala, T., Saramäki, J., and Mubita, O. (1989), "Management Planning System for Tree Plantations. A Case Study for Pinus Kesiya in Zambia", Working Paper, The Finnish Forest Research Institute, Joensuu, Finland.

Sprague, R. H. and Carlson, E. C. (1982), Building Effective Decision Support Systems, Prentice-Hall.

Steuer, R. (1986), Multiple Criteria Optimization: Theory, Computation, and Application, John Wiley & Sons, New York.

Van Hee, K. M. (1987), "Features of the Architecture of Decision Support Systems", Paper presented at the 12th Symposium on Operations Research, Gesellschaft fur Mathematik, Ökonomie und Operations Research, Universität Passau, September 9-11.

Wallenius, H., Leschine, T. M. and Verdini, W. (1987), "Multiple Criteria Decision Methods in Formulating Marine Pollution Policy: A Comparative Investigation", unpublished manuscript.

Youber, E. S. (1986), "Systems to Think With: A Response to "A Vision for Decision Support Systems"", Journal of Management Information Systems II, 85-97.

Wierzbicki, A. (1980), "The Use of Reference Objectives in Multiobjective Optimization", in G. Fandel and T. Gal (Eds.), Multiple Criteria Decision Making, Theory and Application, Springer-Verlag, Berlin, 468-486.

Wierzbicki, A. (1986), "On the Completeness and Constructiveness of Parametric Characterizations to Vector Optimization Problems", OR-Spectrum 8, 73-87.

CISM-IIASA   August 29, 1990
Summer School
Pekka Korhonen

EXERCISE #1

Suppose that DM tries to find the "best" product-mix for three products: Product 1, Product 2, and Product 3. The production of these products requires the use of one machine (Mach.Hours), man-power (Man Hours), and two critical materials (Crit.Mat 1 and Crit.Mat 2). Selling the products results in a profit. The experts have estimated the coefficient matrix for the problem (Table 1):

	Product 1	Product 2	Product 3
Mach.Hours	1.5	1.0	1.6
Man Hours	1.0	2.0	1.0
Crit.Mat 1	9.0	19.5	7.5
Crit.Mat 2	7.0	20.0	9.0
Profit	4.0	5.0	3.0

TABLE 1: The coefficient matrix of the model

The first column in Table 1 tells that to produce one unit of Product 1 requires 1.5 hours the capacity of the machine needed, 1.0 man-hour, 9 units of critical material 1, and 7 units of critical material 2. To sell one unit of Product 1 produce 4 money units. The corresponding information is given in the other columns.

Moreover, suppose that the DM is not able to specify precisely, what "the best" product-mix means. To an expert he/she describe his/her problem as follows:

"I would, of corse, like to make as much profit as possible, but because it is difficult to obtain certain critical materials, I would like to use them as little as possible (presently I have 96 units of them in storage). Only one machine is used to produce the products. The machine operates without any problems for at least 9 hours, but I know from experience that it is likely to break down if used for more than 12 hours. The length of the regular working day is 10 hours. People are willing to work overtime, but it is costly and they are tired the next day. Therefore, if possible, I would like to avoid it. Finally, product 3 is very important to a major customer, and I cannot totally exclude it from the production plan."

Assume that your duty is to help the DM find a proper solution.

♦   How do you help him/her to find a satisfactory solution?

♦   Can you immediately, in the beginning, specify his/her objectives and constraints. If yes, what they are?

♦   Are you sure that the solution you supposed is nondominated? In which sense?

♦   Do you think that the DM is convinced with the solution you helped him/her to find?

EXERCISE #2

Assume that you want to invest 10,000 dollars profitably. Assume further that you have identified four relatively good investment options. It is possible to invest the money in **one or several options**. However, they are not riskless, and the returns depend on the general state of the economy (Declining, Stable, Improving). The experts have given the following table for your help:

Economy\ Options:	1	2	3	4
Declining	-2	4	-7	15
Stable	5	3	9	4
Improving	3	0	10	-8

TABLE 2: Return per unit (%)

Nobody knows the future state of the economy in advance.

♦   How do you invest your 10,000 dollars most profitably?

♦   Are you sure that your decision alternative is nondominated?

DECISION SUPPORT SYSTEMS AND MULTIPLE-CRITERIA OPTIMIZATION

A. Lewandowski
Wayne State University, Detroit, USA

1 Concepts of decision support systems

The concept of a decision support system, though quite widely used and developed in applied research, is by no means well defined. However, without attempting to give a restrictive definition it is necessary to review main functions and various types of decision support systems.

The main function of such systems is to *support decisions made by humans.* It contrasts to decision automation systems that replace humans in repetitive decisions because these decision problems are either too tedious or require very fast reaction time or very high precision. In this sense, every information processing system has some functions of decision support. However, modern decision support systems concentrate on the functions of *helping human decision makers in achieving better decisions.*

This statement is too general to be treated as operational definition of Decision Support System. According to this "definition" *everything* what helps a decision maker in the process of decision making should be classified as DSS – including comfortable office, efficient secretary and a cup of coffee. Clearly, more precise definition of Decision Support System is necessary.

Unfortunately, there is no consensus regarding the definition of Decision Support Systems. Usually, various authors speak about *characteristics* of decision support systems, rather then strict definitions. Moreover, it seems to be evident that the issue of defining the notion of Decision Support System must have the operational character: the same software tools can be used in various ways – either for supporting decision or for other

---

*Paper to be presented on the Summer School on Decision Support Systems organized jointly by the International Center for Mechanical Sciences (CISM), Udine, Italy, University of Udine and the International Institute for Applied System Analysis (IIASA), Laxenburg, Austria. This paper summarizes research performed by the author and the team of Polish scientists led by A.P. Wierzbicki, within the research agreement between the System and Decision Sciences Program of IIASA and the Polish Academy of Sciences. The complete results of this research are presented in IIASA publications WP–88–03 and WP–88–71 *and other* IIASA publications. Review of the DSS software products is presented in the paper WP–88–109 where other references can be found.

purposes. The typical example is the spreadsheet program. There are several ways of using this program (e.g. Lotus 1-2-3) for providing quite efficient decision support, if a suitable collection of macros is developed by the user (see Carlsson and Stabel, 1986, Jones, 1986). In other situations the same program can be used like a data base, just to provide some statistical characteristics of historical data. Therefore, the question, whether *a spreadsheet is a Decision Support System* has not too much sense. The answer depends on the user applying this program for decision analysis. The same applies to the question *is the expert system a Decision Support System?* Clearly, the answer depends on the way how the expert system is used.

However, despite of all semantic problems it is possible to list several basic characteristics of Decision Support Systems. Following Parker and Al-Utabi (1986), Keen and Scott-Morton (1978) and Sprague and Carlson (1982) we can state as follows:

- Decision Support Systems assist managers in their decision processes in semi–structured tasks,

- they support and enhance rather than replace managerial judgment,

- they improve the effectiveness of decision making rather than its efficiency,

- they attempt to combine the use of models or analytical techniques with traditional data access and retrieval functions,

- they specifically focus on features which make them easy to use by non-computer people in an interactive mode,

- they emphasize flexibility and adaptability to accommodate changes in the environment and the decision making approach of the user.

The notion of Decision Support System has been recently criticized by some authors what triggered a broad discussion regarding basic principles of Decision Support Systems. We will not discuss all arguments formulates as a product of this discussion (see Naylor, 1982). This discussion has resulted in improving of understanding the notion of Decision Support System.

The most complete analysis of the notion of Decision Support System has been recently done by Thierauf (1982, 1988). He provides a careful study of all aspects of decision process, components of decision situation as well as possible approaches for solving decision problems. It is not possible to quote here all his arguments and all his conclusions. It is necessary, however, to refer to the following important observations:

- DSS technology must be assembled into systems that are compatible with users' managerial style and that provide support the user feels is valuable,

- User–oriented DSS is *not* intended to make decisions for managers but to provide informational, computational, and display support to improve their decision making,

- DSS is designed for a particular set of user application. As a result, there is literally no limit to the variety of DSS applications possible, which makes it quite different from previous management information systems.

Concluding, Thierauf gives his own definition of Decision Support System:

> *Decision support system allows the decision maker to combine personal judgment with computer output in a user-machine interface to produce meaningful information for support in the decision-making process. Such systems are capable of solving all types of problems (structured, semistructured and unstructured) and use query capabilities to obtain information by request. As deemed appropriate, they use quantitative models as well as database elements for problem solving. From an enlarged perspective, decision support systems are an integral part of the decision maker's approach to problem finding, which stresses a broad view of the organization by employing the "management by perception principle"*

Being more detailed, he specifies a set of *characteristics* of Decision Support Systems. Most of these characteristics do not exist in the prior information systems. The following is the list of characteristics provided by Thierauf:

Problem Finding and Problem Solving

- *A broad approach to support decision making with an accent on "management by perception".* Decision Support Systems go beyond capabilities of information systems by taking a broad view of the organization in terms of supporting decisions. They use "management by perception" whereby, managers are assisted in perceiving important future trends and helped in adapting the organization to upcoming conditions. This is opposed to the "management by exception" principle used in the previous information systems.

- *User-machine interface, which permits the user to retain control through the problem-finding and problem-solving processes.* The user has the capability to retrieve, manipulate, present and store data such that there is a human-machine dialogue. The user has complete control over all stages of either problem finding or problem solving.

- *User support in solving well-structured, semistructured and unstructured problems.* Although the focus is on the answers to semistructured and unstructured problems, well-structured problems can also be solved in a DSS environment. DSS recognizes the need for bringing together human judgment and computerized information for improving the quality of the final decision in either problem finding or problem solving.

- *Use of quantitative models.* Based on the needs of the problem, one or more mathematical and/or statistical models are employed to assist the user in evaluating alternative solutions. The real payoff from quantitative models comes from integrating them into the decision support system as decision tool.

- *Use of financial planning languages, simulation languages, statistical packages and other modelling tools.* In conjunction with the use of standard and custom–made quantitative models, there is a great accent today on using more advanced software tools to solve a wide range of organizational problems in an interactive processing mode.

Interactive Processing Mode

- *Query capabilities to obtain information by request.* In DSS, query capabilities go beyond interactive computation and include responsiveness, which implies using the system as an extension of the individual's reasoning process through the decision making process.

- *Use of management workstation.* Management workstations must allow user to perform a wide range of computer functions in an interactive processing mode. The more sophisticated management workstations offer data processing, color graphics, modelling languages, word processing capabilities.

- *Convenient and easy to use approach.* The hallmark of effective DSS is that it is easy to use; not only does it assist the user in supporting decisions via human–machine interface, but it also allows the user to pursue his own natural tendencies to problem finding and problem solving.

- *Adaptivity over time.* The user is able to confront changing conditions and adapt the system to meet these changes. The time factor for effecting system changes may range from a few days to several month.

Comprehensive System Approach

- *Integrated systems of functional areas.* Data for use by all systems are processed along broader, functional lines rather than the traditional, narrow departmental lines. Integrated systems and subsystems allow managers and their personnel to retrieve and manipulate information of concern to them for supporting decisions.

- *Enlarged database, with integration of external and internal data elements.* Contents of the database for DSS must go beyond just providing historical information about current and past operations. It must also contain appropriate external information which is compatible with internal information contained in the database. Generally, it is desirable to use a database management system to assist in a user–machine dialogue.

- *Output directed to organization personnel at all levels.* Although DSS has the capability to supply top and middle management with important short- to long-range planning information for decision making that was not available with earlier computer systems, it is also capable of providing lower management and operating personnel with the necessary output for supporting decisions on controlling current operations. A comprehensive DSS approach assists in supporting all organizational members in decision making where deemed necessary.

Better understanding of the *function* of Decision Support System requites more deep understanding of the *decision process* itself.

Despite the fact, that a very general definition cannot be formulated, it is possible to think about Decision Support Systems supporting particular classes of problems or for building tools which as stand-alone products cannot be considered as DSS but can be used for building such systems. Usually, concentration on a given application domain or on a given technology allows to say much more about the notion of DSS. The example of such an approach has the recent paper by Van Hee (1989) who presents a framework and formalization of DSS for supporting scheduling problems. To the same category belongs research by Anthonisse at al. (1989) who investigate general properties of *interactive planning systems.* Trying to follow this direction, we will concentrate on systems having the following properties:

- The integral part of the system (despite the lack of definition, we will call these systems Decision Support System) is a *mathematical model* which can be used for predicting the consequences of decisions proposed either by decision maker, or by the system.

- Decision making has a qualitative character, i.e. different decisions can be compared and for this comparison the decision maker or the system can utilize *qualitative performance factors (criteria).*

It is possible to invent different approaches to the issue of clarification what is and what is not a DSS. Namely, we can start not from a very general investigation of this notion, applying the top–down approach, but instead apply the bottom–up way. To do this we can ask the following question: *in what way the tools, methodologies and framework being in our disposal can support decision making processes?* Although this approach sometimes is strongly criticizes (called "toolism") it seems that it can be applicable in the area of DSS. For well defined tools we can give a precise definition of a *decision environment* for which these tools can provide support. Following this idea, in the sequel we will speak about *model based, multiple–criteria Decision Support Systems* trying to figure out possible range of application of such systems.

2   The Quasisatisficing Decision Framework

To develop the framework for Decision Support Systems it is necessary to specify the concept of *rational decision*. This is beyond the scope of this paper to discuss all possible aspects of rationality. Detailed discussion of the problem can be found in the paper by Lewandowski and Wierzbicki (1988a).

There exist several frameworks for analytical rationality. We can represent them best when assuming a certain mathematical structure of the decision situation. Such a structure consists of:

- a space of decisions (alternatives, options, controls, designs etc.) denoted by E_x; if this space is a discrete set, we speak about discrete alternatives,

- a constraint set of admissible decisions $X_o \subseteq E_x$,

- a space of outcomes (attributes, objective outcomes, objectives, performance indices, etc.) denoted by E_y,

- an outcome mapping $f : E_x \to E_y$, which also defines the set of attainable outcomes $Y_o = f(X_o) \subset E_y$; this mapping might be given explicitly by a *substantive model of the decision situation* or be supplied judgmentally by experts evaluating alternatives along various attributes, in which case we have *judgmental model evaluation*.

- a partial preordering in the space of outcomes that is usually implied by the decision problem and usually has some obvious interpretation, such as maximization of profit competing with the maximization of market share, etc. A standard assumption is that this preordering is *transitive* and can be expressed by a *positive cone* $D \subset E_x$.

- a complete preordering in the space of outcomes or, at least, in the set of attainable outcomes, which is usually not given in any precise mathematical form, but is contained in the mind of the decision maker, such as how actually the preferences between the maximization of profit and the maximization of market share should be distributed in the above example.

The main differences between various frameworks of rationality that lead to diverse approaches to interactive decision support are concerned with the assumptions about this complete preordering and the way of its utilization in the DSS. This issue is also closely related with the way in which the DSS interacts with the decision maker. Some variants of DSS require that the user answers enough questions for an adequate estimation of this complete preordering, some other variants need only general assumptions about the preordering, still other variants admit a broad interpretation of this preordering and diverse frameworks of rationality that might be followed by the user.

The most strongly established rationality framework is based on the assumption of *maximization of a value function or an utility function*. Under rather general assumptions, the complete preordering that represents the preferences of the decision maker can be represented by an utility function $u : E_y \rightarrow R^1$ such that by maximizing this function over $x \in X_o$ we can select the decision which is most preferable to the decision maker. The publications related to this framework are very numerous, for example, the book by Keeney and Raiffa (1976).

There are many fundamental and technical difficulties related to the identification of such utility function. Leaving aside various technical difficulties, we should stress the fundamental ones.

- *Firstly*, a continuous utility function only exists if there is no strict hierarchy of values between decision outcomes, that is if all decision outcomes can be aggregated into one value. This does not mean that hierarchical dependence between outcomes cannot be incorporated in this framework, but such dependencies must be treated as constraints and cannot be evaluated in the decision process. Thus, the utility maximization framework represents *the culture of an entrepreneur facing an infinite market* which, although it represents the behavior of many human decision makers, is by no means the universal case of human rationality (Rapoport, 1984). Rapoport states the following:

   "...The difficulty is that in many instances a utility function satisfying certain apparently innocuous consistency criteria cannot be established to begin with. The investigator who is interested not in actor's utility function per se but rather in his decision behavior is left by no choice but to by-pass the utility problem altogether and work with actual pay-offs used in the experiments, for instance, money..."

- *Secondly*, while the utility maximization framework might be a good predictor of mass economic phenomena, *it has many drawbacks as a predictor of individual behavior* – see, e.g. Fisher (1979), Erlandson (1981). According to the results of research presented in these papers, the utility function approach can be used in a rather simple, laboratory environment, but can fail in more complex situations.

- *Thirdly* – and most importantly for applications in decision support systems – an experimental identification and estimation of an utility function requires many questions and answers in the interaction with the decision maker (e.g. Keeney and Sicherman, 1975). Users of decision support systems are typically not prepared to answer that many questions, for several reasons. They do not like to waste time and they do not like to disclose their preferences in too much detail. This is because they intuitively perceive that the decision system should support them in learning about the decision situation and thus they should preserve the right to change their minds and preferences. Therefore, if any approximation of an utility function is used in a decision support system, it should be *nonstationary in time in order to account for the learning and adaptive nature of a decision making process.* Such an approximation cannot be very detailed, it must have a reasonably simple form characterized by some adaptive parameters that can aggregate the effects of learning.

Another rationality framework, called *satisficing decision making*, was formulated by Simon (1969) and further extended by many researchers (Erlandson, 1981). Originally, this approach assumes that human decision makers do not optimize because of the difficulty of optimization operations, uncertainty of typical decision environment, and complexity of the decision situations in large organizations. Therefore, this approach was sometimes termed *bounded rationality*. The recent article by March (1986) gives a very detailed analysis of this concept as well as other possible views of rationality.

A very important contribution of the satisficing framework is the observation that decision makers often use *aspiration levels* for various outcomes of decisions (see Tietz, 1983). In the classical interpretation of the satisficing framework, these aspiration levels indicate when to stop optimizing. While more modern interpretations might prefer other rules for stopping optimization, the concept of aspiration levels is extremely useful for aggregating the results of learning by the decision maker:

> *aspiration levels represent values of decision outcomes that can be accepted as reasonable or satisfactory by the decision maker and thus are aggregated, adaptable parameters that are sufficient for a simple representation of his accumulated experience.*

In order to develop a broader framework that would be useful for decision support for decision makers representing various perspectives of rationality, Wierzbicki (1982, 1984, 1986) proposed the following *principles of quasisatisficing decision making.*

A *quasisatisficing decision situation* consists of:

- one or several *decision makers* or *users* that might represent any perspective of rationality and have the right of changing their minds due to learning. They have also the right of stopping optimization for any reason,

- a decision support system that might be either fully computerized or include also human experts, analysts or advisors.

It is assumed that:

- The user evaluates possible decisions on the basis of a vector of attributes or objective outcomes. These factors can be expressed in numerical scale (quantitatively) or in verbal scale (qualitatively), like *bad, good* or *excellent*. Each factor can be additionally constrained by specifying special requirements on it that must be satisfied. Beside this, objective outcomes can be characterized by their type: maximized, minimized, stabilized – that is, kept close to a given level (which corresponds to foregoing optimization), or floating – that is, included for the purpose of additional information or for specifying constraints. The user has the control over the specification of objective outcomes together with their types and the possible aggregation of such factors.

- One of the basic means of communication of the user with the decision support system is his specification of aspiration levels for each objective outcome; these aspiration levels are interpreted as reasonable values of objective outcomes. In more complex situations, the user can specify two levels for each objective outcome – an *aspiration level* interpreted as above and a *reservation level* interpreted as the lowest acceptable level for the given objective outcome.

- Given the information specified by the user the decision support system following the quasisatisficing principle should use this information, together with other information contained in the system, in order to propose to the user one or several alternative decisions that are best attuned to this guiding information. When preparing (generating or selecting) such alternative decisions, the decision support system should not impose on the user the optimizing or the satisficing or any other behavior, but should follow the behavior that is indicated by the types of objective outcomes. This means that the decision support system should optimize when at least one objective outcome is specified as minimized or maximized and should suffice (stop optimizing upon reaching aspiration levels) when all objective outcomes are specified as stabilized.

In order to illustrate possible responses of a quasisatisficing decision support system to the information given by the user, let us assume that all specified objective outcomes are supposed to be maximized and have specified aspiration levels. We can then distinguish the following cases:

- *Case 1*: the user has overestimated the possibilities implied by admissible decisions and there is no admissible decision such that the values of all objective outcomes are exactly equal to their aspiration levels. In this case, however, it is possible to propose a decision for which the values of objective outcomes are as close as possible to their aspiration levels. The decision support system should tentatively propose one decision or several such decisions to the user.

- *Case 2*: the user underestimated the possibilities implied by admissible decisions and there exist a decision which results in the values of objective outcomes exactly equal to the specified aspiration levels. In this case, it is possible to propose a decision which improves all objective outcomes uniformly as much as possible. The decision support system should inform the user about this case and tentatively propose one decision or several such decisions.

- *Case 3*: the user, by a chance or as a result of a learning process, has specified aspiration levels that are uniquely attainable by an admissible decision. The decision support system should inform the user about this case and specify the details of the decision that result in the attainment of aspiration levels.

In the process of quasisatisficing decision support, all aspiration levels and the corresponding decisions proposed by the system have tentative character. If a decision proposed by the system is not satisfactory to the user, he can modify the aspiration levels and obtain new proposed decisions, or even modify the specification of objective outcomes or constraints. The process is repeated until the user learns enough to make the actual decision himself or to accept a decision proposed by the system.

The process of quasisatisficing decision making can be formalized mathematically (Wierzbicki, 1986, Lewandowski and Wierzbicki, 1988a) and the mathematical formalization can be interpreted in various ways. Let us consider an interpretation that corresponds to the framework of utility maximization. We assume that the user has a nonstationary utility function that changes in time due to his learning about a given decision situation. At each time instant, however, he can intuitively and tentatively (possibly with errors concerning various aspects of the decision situation) maximize his utility. This tentative maximization determine his aspiration levels, denoted here by $w \in E_y$.

When the decision maker communicates the aspiration levels w to the decision support system, the system should use this information, together with the specification of the decision situation, in order to construct an approximation of his utility function that is relatively simple and easily adaptable to the changes of aspiration levels, treated as parameters of this approximation. By maximizing such an approximative utility function while using more precise information about the attainability of alternative decisions and other aspects of the decision situation – for example, expressed by the substantive model of the decision situation incorporated by expert advice into the decision support system – a tentative decision can be proposed to the user.

Such a tentative approximation of the user's utility function, constructed in the decision support system only in order to propose a tentative decision to the learning decision maker, is called here *order–consistent achievement function* or simply *achievement function* and has the form $u(y) = s(y, w)$. By an order consistent achievement function we understand here either an order representing or an order approximating achievement function, according to the following definitions:

An *order representing achievement function* is a continuous function $s : Y_o \times E_y \to R^1$, with arguments $y \in Y_o$ and $w \in E_y$ interpreted as an attainable objective outcome vector and an aspiration level vector that satisfies the following requirements:

- It is *strictly order preserving (monotone)* with respect to y and the positive cone D implied by the partial preordering (according to the types of objective outcomes) specified by the decision maker, that is, for all $w \in E_y$:

$$y_2 - y_1 \in \text{int} D \Rightarrow s(y_1, w) < s(y_2, w) \qquad (1)$$

- It is *order representing* with respect to y and the positive cone D, that is, for all $w \in E_y$:

$$\{y \in E_y : s(y,w) \geq 0\} = w + D \tag{2}$$

If $E_y = R^m$ and all objective outcomes are maximized, $D = R_+^m$, then a simple example of an order representing achievement function is:

$$s(y,w) = \min_{1 \leq i \leq m} \frac{y_i - w_i}{a_i} \tag{3}$$

where a_i represent some scaling units for subsequent objectives. Because of these scaling units, this function has a cardinal form i.e. does not depend on positive affine transformations of the space of outcomes together with scaling units.

An *order–approximating achievement function* is a continuous function $s : Y_o \times E_y \rightarrow R^1$, with arguments $y \in Y_o$ and $w \in E_y$ interpreted as an attainable objective outcome vector and an aspiration level vector that satisfies the following requirements:

- It is *strongly order preserving (monotone)* with respect to y and the positive cone D, that is, for all $w \in E_y$:

$$y_2 - y_1 \in D \setminus (D \cap -D) \Rightarrow s(y_1, w) < s(y_2, w) \tag{4}$$

- It is *order approximating* with respect to y and the positive cone D, that is, for all $w \in E_y$ and for some small $\varepsilon > 0$:

$$w + D \subset \{y \in E_y : s(y,w) \geq 0\} \subseteq w + D_\varepsilon \tag{5}$$

where

$$D_\varepsilon = \{y \in E_y : \text{dist}(y, D) < \varepsilon \parallel y \parallel\} \tag{6}$$

If $E_y = R^m$ and $D = R_+^m$, then a simple example of an order approximating achievement function is:

$$s(y,w) = \max_{1 \leq i \leq m} \frac{y_i - w_i}{a_i} + \frac{\varepsilon}{m} \sum_{i=1}^{m} \frac{y_i - w_i}{a_i} \tag{7}$$

Intuitively speaking, we might say that if $w \in Y_o - D$, then the maximization of $s(y,w)$ over $y \in Y_o$ represents a uniform maximization of all components of the surplus $y - w \in D$; if $w \notin E_y - D$, then the same maximization represents distance minimization between the sets $w + D$ and

$$\{y \in Y_o : Y_o \cap (y + D \setminus (D \cap -D)) = \emptyset\} \tag{8}$$

which is the set of *generalized Pareto optimal objective outcomes in the sense implied by the positive cone* D.

Important properties of order consistent achievement functions are summarized by the following theorems (Wierzbicki, 1986):

Theorem 1 *If $s(y,w)$ is strongly order preserving then its maximal points in $y \in Y_o$ are generalized Pareto optimal, that is, satisfy the following condition:*

$$\tilde{y} = \arg\max_{y \in Y_o} s(y,w) \Rightarrow Y_o \cap (\tilde{y} + \tilde{D}) = \emptyset \qquad (9)$$

where

$$\tilde{D} = D \setminus (D \cap -D) \qquad (10)$$

If $s(y,w)$ is strictly order preserving then its maximal points in $y \in Y_o$ are generalized weakly Pareto optimal, that is, satisfy the following condition:

$$\tilde{y} = \arg\max_{y \in Y_o} s(y,w) \Rightarrow Y_o \cap (\tilde{y} - \text{int}D) = \emptyset \qquad (11)$$

Theorem 2 *If $s(y,w)$ is order approximating and $w \in Y_o$ is generalized properly Pareto optimal (with trade off coefficients bounded by ε and $1/\varepsilon$), then the maximum of $s(y,w)$ in $y \in Y_o$, equal zero, is attained at $y = w$. If $s(y,w)$ is order representing and $w \in Y_o$ is generalized weakly Pareto optimal, then the maximum of $s(y,w)$ in $y \in Y_o$, equal zero, is attained at $y = w$.*

Thus, the usefulness of achievement functions in building interactive decision support systems follows from the following properties:

- maximization of an order approximating achievement function results in Pareto optimality, no matter whether the aspiration level is attainable or not. Order representing functions are less useful, because their maxima are only weakly Pareto optimal,

- if a decision $\hat{x} \in X_o$ and the corresponding objective outcome $\hat{y} \in Y_o$ maximize an order approximating achievement function and $\hat{s} = s(\hat{y}, w) = 0$ then the aspiration levels w are attainable and Pareto optimal,

- if, in the above situation, $\hat{s} < 0$, then the aspiration levels w are not attainable,

- if, in the above situation, $\hat{s} > 0$, then the aspiration levels w are attainable, but not Pareto optimal.

Therefore, an order approximating achievement function can be used for computing Pareto optimal decisions as well as for checking for Pareto optimality and attainability of an arbitrarily given $w \in E_y$. Moreover, the value of such achievement function can be meaningfully interpreted – it can be treated as a *qualitative distance* between a given decision $\hat{x}$ or its objective outcome $\hat{y}$ and the aspiration level w.

Beside the achievement functions specified by equations (3) and (7), there are many other forms of this function (Wierzbicki, 1986). Another example of an order representing achievement function might be:

$$s(y, w) = \max \left\{ \varrho \max_{i \leq i \leq m} \frac{y_i - w_i}{a_i}, \quad \frac{1}{m} \sum_{i=1}^{m} \frac{y_i - w_i}{a_i} \right\} \tag{12}$$

The above function is especially useful when applied to decision support systems with substantive models of linear multiobjective optimization type, when its maximization can be reduced by suitable transformation of variables to a single objective linear programming problem with additional constraints (Lewandowski at al., 1985, 1987, Lewandowski and Wierzbicki, 1988c).

Practical experiments with this approach (Lewandowski at al., 1985, Dobrowolski and Zebrowski, 1987) have shown that the *language of aspiration levels* coincides very well with the style of thinking of practical decision makers. The information which is required from the user is easy to express, as opposed to other approaches based on pairwise comparisons, explicit weighting factors, estimation of other forms of utility functions, etc.

Theoretically, the learning process of interaction with a quasisatisficing decision support system via changing aspiration levels might not be sufficient for all decision makers: some of them might learn sufficiently to select their preferred decision, some others might still be puzzled and require some help in the convergence to their best preferred decision. There are several ways of organizing such support for the user in changing his aspiration levels that the corresponding maxima of achievement functions converge to the maximum of his utility function. One way consists in the visual interactive approach of Korhonen (1988) and Korhonen and Laakso (1986), or *directional scanning* of aspirations and the corresponding maxima of achievement functions.

3   Decision support systems of DIDAS family

A typical procedure of working with a system of DIDAS family consists of several phases:

- *Phase A.* The definition and edition of a model of analysed process and decision situation by analyst(s)

- *Phase B.* The definition of the multiobjective decision problem using the model, by the final user (the decision maker) together with analyst(s)

- *Phase C.* The initial analysis of the multiobjective decision problem, resulting in determining bounds on efficient outcomes and, possibly, a neutral efficient solution and outcome, by the user helped by the system

- *Phase D.* The main phase of interactive, learning review of efficient solutions and outcomes for the multiobjective decision problem, by the user helped by the system

- *Phase E.* An additional phase of sensitivity analysis (typically, helpful to the user) and/or convergence to the most preferred solution (typically, helpful only to users that adhere to utility maximization framework).

These phases have been implemented differently in various systems of DIDAS family; however, we describe them here comprehensively.

Phase A: Model definition and edition

There are four basic classes of models that have been used in various systems of DIDAS family:

- multiobjective linear programming models,
- multiobjective dynamic linear programming models,
- multiobjective nonlinear programming models,
- multiobjective dynamic nonlinear programming models.

Linear programming model

A multiobjective linear programming model consists of the specification of vectors of n decision variables $x \in R^n$, m outcome variables $y \in R^m$, linear model equations defining the relations between the decision variables and the outcome variables with model bounds defining the lower and upper bounds for all decision and outcome variables:

$$y = Ax \qquad x^{lo} \leq x \leq x^{up} \qquad y^{lo} \leq y \leq y^{up} \tag{13}$$

In the above formula A is a $m \times n$ matrix of coefficients. Between outcome variables, some might be chosen as guided outcomes corresponding to equality constraints. We will denote these variables by $y^c \in R^{m'} \subset R^m$ and the constraining value for them by b^c to write the additional constraints in the form:

$$y^c = A^c x = b^c \qquad y^{c,lo} \leq b \leq y^{c,up} \tag{14}$$

where A^c is the corresponding submatrix of A. Other outcome variables can be chosen as optimized objectives or objective outcomes. Some of the objective variables might be originally not represented as outcomes of the model, but they can be always added by modifying this model. In any case, the corresponding objective equations in linear models have the form:

$$q = Cx \tag{15}$$

where C is another submatrix of A. Thus, the set of attainable objective outcomes is $Q_0 = CX_0$ and the set of admissible decisions X_0 is defined by:

$$X_0 = \{x \in R^n : \quad x^{lo} \leq x \leq x^{up} \quad y^{lo} \leq Ax \leq y^{up} \quad A^c x = b^c\} \tag{16}$$

By introducing additional variables and constraints, the problem of maximizing functions like (12) over outcomes (15) and admissible decisions (16) can be equivalently rewritten to a parametric linear programming problem, with the leading parameter $\bar{q}$. Therefore in next phases of the process a linear programming algorithm can be applied.

Nonlinear model

A multiobjective nonlinear programming model consists of the specification of vectors of n decision variables $x \in R^n$ and of m outcome variables $y \in R^m$ together with nonlinear model equations defining the relations between the decision variables and the outcome variables and with model bounds defining the lower and upper bounds for all decision and outcome variables

$$y = g(x) \qquad x^{lo} \le x \le x^{up} \qquad y^{lo} \le y \le y^{up} \tag{17}$$

where $g : R^n \rightarrow R^m$ is a (differentiable) function. In fact, the user or the analyst does not have to define the function g explicitly; he can also define it recursively determining further components of this vector-valued function as functions of formerly defined components. Between outcome variables, some might be chosen as guided outcomes corresponding to equality constraints. We will denote these variables by $y^c \in R^{m'} \subset R^m$ and the constraining' value for them by b^c to write the additional constraints in the form:

$$y^c = g^c(x) = b^c \qquad y^{c,lo} \le b^c \le y^{c,up} \tag{18}$$

where g^c is a function composed of corresponding components of g. In phase B, other outcome variables can be also chosen as optimized objectives or objective outcomes. The corresponding objective equations have the form:

$$q = f(x) \tag{19}$$

where f is also composed of corresponding components of g. Thus, the set of attainable objective outcomes is $Q_0 = f(X_0)$ where the set of admissible decisions X_0 is defined by:

$$X_0 = \{x \in R^n : \quad x^{lo} \le x \le x^{up} \quad y^{lo} \le g(x) \le y^{up} \quad g^c(x) = b^c\} \tag{20}$$

In further phases of working with nonlinear models, an order-approximating achievement function must be maximized. Since this maximization is performed repetitively, there are special requirements for the solver: it should be robust, adaptable and efficient. The requirement for robustness means that it should compute reasonably fast an optimal solution for optimization problems of a broad class of functions $g(x)$ and $f(x)$) without requiring from the user to adjusts parameters of the algorithm in order to obtain a solution. The experience in applying nonlinear optimization algorithms in decision support systems (Kreglewski and Lewandowski, 1983, Kaden and Kreglewski, 1986) has led to the choice of an algorithm based on penalty shifting technique and projected conjugate gradient method. Since a penalty shifting technique anyway approximates nonlinear constraints by penalty terms, an appropriate form of an achievement function that differentiably approximates function (12) has been also developed and is actually used. This *smooth order-approximating achievement function* has the form:

$$s(q, \bar{q}) = 1 - \left\{ \frac{1}{p} \left[\sum_{i=1}^{p''} (w_i)^\alpha + \sum_{i=p''}^{p+1} \max((w_i', w_i''))^\alpha \right] \right\}^{1/\alpha} \qquad (21)$$

where w_i, w_i', w_i'' are functions of q_i, $\bar{q}_i$:

$$w_i(q_i, \bar{q}_i) = \begin{cases} \frac{q_{i,max} - q_i}{s_i}, & \text{if } 1 \leq i \leq p' \\ \frac{q_i - q_{i,min}}{s_i}, & \text{if } p' + 1 \leq i \leq p'' \end{cases} \qquad (22)$$

$$\left. \begin{aligned} w_i'(q_i, \bar{q}_i) &= \frac{q_{i,max} - q_i}{s_i'} \\ w_i''(q_i, \bar{q}_i) &= \frac{q_i - q_{i,min}}{s_i''} \end{aligned} \right\}, \quad \text{if } p'' + 1 \leq i \leq p, \qquad (23)$$

and the dependence on $\bar{q}_i$ results from a special definition of the scaling units that are determined by:

$$s_i = \begin{cases} q_{i,max} - \bar{q}_i, & \text{if } 1 \leq i \leq p', \\ \bar{q}_i - q_{i,min}, & \text{if } p' + 1 \leq i \leq p'', \end{cases} \qquad (24)$$

$$\left. \begin{aligned} s_i' &= q_{i,max} - \bar{q}_i \\ s_i'' &= \bar{q}_i - q_{i,min} \end{aligned} \right\}, \quad \text{if } p'' + 1 \leq i \leq p \qquad (25)$$

In the initial phase of the analysis the values $q_{i,max}$ and $q_{i,min}$ are set to the upper and lower bounds specified by the user for the corresponding outcome variables. In next steps of the procedure they are modified. The parameter $\alpha \geq 2$ is responsible for the approximation of the function (12) by the function (21): if $\alpha \to \infty$ and $\epsilon \to 0$, then these functions converge to each other. However, the use of too large parameters results in badly conditioned problems when maximizing function (21), hence $\alpha = 4, \ldots, 8$ are suggested to be used.

The function (21) must be maximized with $q = f(x)$ over $x \in X_0$, while X_0 is determined by simple bounds $x^{lo} \leq x \leq x^{up}$ as well as by inequality constraints $y^{lo} \leq g(x) \leq y^{up}$ and equality constraints $g^c(x) = b^c$. In the shifted penalty technique, the following function is minimized instead:

$$P(x, \xi', \xi'', \xi, u', u'', v) = -s(f(x), \bar{q}) + \frac{1}{2} \sum_{i=1}^{p'} \xi_i'(\max(0, g_i(x) - y_i^{up} + u_i'))^2 +$$

$$+ \frac{1}{2} \sum_{i=p'}^{p''+1} \xi_i''(\max(0, y_i^{lo} - g_i(x) + u_i''))^2 + \frac{1}{2} \sum_{i=p''}^{p+1} \xi(g_i^c(x) - b_i^c + v_i))^2 \qquad (26)$$

where ξ', ξ'', ξ are penalty coefficients and u', u'', v are penalty shifts. This function is minimized over x such that $x^{lo} \leq x \leq x^{up}$ applying conjugate gradient directions, projected on these simple bounds if one of the bounds becomes active. When a minimum of this penalty function with given penalty coefficients and given penalty shifts is found, the following steps are undertaken:

- the violations of all outcome constraints are computed,

- the penalty shifts and coefficients are modified according to the shifted-increased penalty technique (Wierzbicki, 1984b),

- the penalty function is minimized.

This procedure is repeated again until the violations of outcome constraints are admissibly small. The results are equivalent to the outcomes obtained by maximizing the achievement function (21) under all constraints. This technique is according to existing experience one of the most robust nonlinear optimization methods.

The description of the standards for defining *models of dynamic nonlinear programming type* will be not presented here, since that can be obtained by combining the previous cases. ⏽

Phase B. The definition of the multiobjective decision analysis problem

For a given model, the user can define various problems of multiobjective analysis by suitably choosing maximized, minimized, stabilized and guided outcomes. In this phase, he can also define which outcomes and decisions should be displayed to him additionally during interaction with the system (such additional variables are called *floating outcomes*). Since the model is typically prepared by an analyst(s) in the phase A and further phases starting with the phase B must be performed by the final user, an essential aspect of all systems of DIDAS family is the user-friendliness of phase B and further phases. This issue has been variously resolved in consequent variants of DIDAS systems. In all these variants, however, the formulation of the achievement function and its optimization is prepared automatically by the system once phase B is completed.

Before the initial analysis phase, the user should also define some reasonable lower and upper bounds for each optimized (maximized, minimized or stabilized) variable, which results in an automatic definition of reasonable scaling units s_i for these variables. In further phases of analysis, these scaling units s_i can be further adjusted.

Phase C. Initial analysis of the multiobjective problem

Once the multiobjective problem is defined, bounds on efficient solutions can be approximated either automatically or on request of the user.

The *upper* bound for efficient solutions could be theoretically obtained through maximizing each objective separately (or minimizing, in case of minimized objectives. In the case of stabilized objectives, the user should know their entire attainable range, hence they should be both maximized and minimized). Jointly, the results of such optimization form a point that approximates from above the set of efficient outcomes $\hat{Q}$, but this point almost never (except in degenerate cases) is in itself an attainable outcome. Therefore, such a point is called the *utopia point* $\hat{q}^{uto}$.

However, this way of computing the upper bound for efficient outcomes is not always practical due to required computational effort. Many systems of DIDAS family use a different way of estimating the utopia point. This way consists in subsequent maximizations of the achievement function $s(q, \bar{q})$ with suitably selected reference points $\bar{q}$. If an objective should be maximized and its maximal value must be estimated, then the corresponding component of the reference point should be very high, while the components of this point for all other maximized objectives should be very low. If an objective should be minimized and its minimal value must be estimated, the corresponding component of the reference point should be very low, while other components of this point are treated as in the previous case. If an objective should be stabilized and both its maximal and minimal values must be estimated, then the achievement function should be maximized twice, first time as if for a maximized objective and the second time as if for a minimized one (while the obtained maximal and minimal values will be denoted by $\hat{q}_i^{uto}$ and $\hat{q}_i^{nad}$, respectively, although it is difficult to say which of them corresponds to the concept of utopia point). Thus, the entire number of optimization runs in utopia point computations is $p'' + 2(p - p'')$. This is especially important in dynamic cases. It can be shown that this procedure gives a very good approximation of the utopia point $\hat{q}^{uto}$.

During all these computations, the lower bound for efficient outcomes can be also estimated. This can be done by recording the lowest efficient outcomes that occur in subsequent optimizations for maximized objectives and the highest ones for minimized objectives. However, such a procedure results in the accurate, tight lower bound for efficient outcomes – called *nadir point* $\hat{q}^{nad}$ – only if $p'' = 2$. For larger numbers of maximized and minimized objectives, this procedure can give misleading results. The accurate computation of the nadir point becomes a very cumbersome computational task requirinng very special approaches (see Isermann and Steuer, 1987).

Therefore, some systems of DIDAS family accept user-supplied estimates of lower bounds for objectives. They also offer an option of improving the estimation of the nadir point in such cases. This option consists in additional p'' maximization runs for achievement function $s(q, \bar{q})$ with the following rules for selecting the reference point if the objective in question should be maximized:

- reference point corresponding to the selected objective $\bar{q}$ is very low,

- reference points are very high for other maximized objectives and very low for other minimized objectives,

- stabilized objectives should be considered as floating.

If the objective in question should be minimized, the similar rules applies: the corresponding reference component should be very high, while other reference components should be treated as in the previous case. By recording the lowest efficient outcomes that occur in subsequent optimizations for maximized objectives and the highest ones for minimized objectives, a better estimation $\hat{q}^{nad}$ of the nadir point is obtained.

For dynamic models, the number of objectives becomes formally very high which would imply a very large number of optimization runs when estimating the utopia point:

$$(p'' + 2(p - p''))(T + 1) \tag{27}$$

Therefore, it is important to obtain approximate bounds on entire trajectories. This can be obtained by $p'' + 2(p - p'')$ optimization runs organized as in the static case, with correspondingly 'very high' and 'very low' reference or aspiration trajectories.

Once the approximate bounds $\hat{q}^{uto}$ and $\hat{q}^{nad}$ are computed and known to the user, they can be utilized in various ways. One way consists in computing a neutral efficient solution, with outcomes situated approximately in the middle of the efficient set. For this purpose, the reference point $\bar{q}$ is situated at the utopia point $\hat{q}^{uto}$ and the scaling units are determined by:

$$s_i = \left| \hat{q}_i^{uto} - \hat{q}_i^{nad} \right|, \qquad 1 \le i \le p \tag{28}$$

for all outcomes. By maximizing the achievement function $s(q, \bar{q})$ with such data, the neutral efficient solution is obtained and can be utilized by the user as a starting point for further interactive analysis of efficient solutions.

Once the utopia and nadir point are estimated and, optionally, a neutral solution computed and communicated to the user, he has enough information about the ranges of outcomes in the problem to start the main interactive analysis phase.

Phase D. Interactive review of efficient solutions and outcomes

In this phase, the user controls the efficient solutions and outcomes computed for him in the system. This can be achieved by changing reference or aspiration points. It is assumed that the user is interested only in efficient solutions and outcomes. Iif he wants to analyse outcomes that are not efficient for the given definition of the problem, he must change the definition which necessitates a repetition of phases B, C.

In the interactive analysis phase, an important consideration is that the user should be able to easily influence the selection of the efficient outcomes $\hat{q}$ by changing the reference point $\bar{q}$ in the maximized achievement function $s(q, \bar{q})$. It can be shown (Wierzbicki, 1986) that best suited for the purpose is the choice of scaling units determined by the difference between the slightly displaced utopia point and the current reference point:

$$s_i = \begin{cases} \hat{q}_i^{uto} - \bar{q}_i + 0.01(\hat{q}_i^{uto} - \hat{q}_i^{nad}) & \text{if } 1 \le i \le p' \\ \bar{q}_i - \hat{q}_i^{uto} + 0.01(\hat{q}_i^{uto} - \hat{q}_i^{nad}) & \text{if } p' + 1 \le i \le p'' \end{cases} \tag{29}$$

for maximized or minimized outcomes. For stabilized outcomes, the scaling units are determined then:

$$\left. \begin{array}{l} s_i' = \hat{q}_i^{uto} - \bar{q}_i + 0.01(\hat{q}_i^{uto} - \hat{q}_i^{nad}) \\ s_i'' = \bar{q}_i - \hat{q}_i^{nad} + 0.01(\hat{q}_i^{uto} - \hat{q}_i^{nad}) \end{array} \right\}, \quad \text{if } p'' + 1 \le i \le p \tag{30}$$

It is assumed now that the user selects the reference components in the range $\hat{q}_i^{nad} \le \bar{q}_i \le \hat{q}_i^{uto}$ for maximized and stabilized outcomes or $\hat{q}_i^{uto} \le \bar{q}_i \le \hat{q}_i^{nad}$ for minimized outcomes. In some DIDAS systems, there is also an option of user-defined weighting coefficients, but the automatic definition of scaling units is sufficient for influencing the selection of efficient outcomes.

The interpretation of the above way of setting scaling units is that the user attaches implicitly more importance to reaching a reference component $\bar{q}_i$ if he places it close to the known utopia component. In such a case, the corresponding scaling unit becomes smaller and the corresponding objective component is weighted stronger in the achievement function $s(q, \bar{q})$. Thus, this way of *scaling relative to utopia-reference difference* is taking into account the implicit information given by the user in the relative position of the reference point. This way of scaling, has been used also by Nakayama and Sawaragi, (1983) and Steuer and Choo (1983).

When the relative scaling is applied, the user can easily obtain by suitably moving reference points, efficient outcomes that are either situated close to the neutral solution, in the middle of efficient outcome set $\hat{Q}_0$, or in some remote parts of the set $\hat{Q}_0$, close to various extreme solutions. Typically, several experiments of computing such efficient outcomes give enough information for the user to select an actual decision. However, there might be some cases in which the user would like to receive further support, either in analysing the sensitivity of a selected efficient outcome, or in converging to some best preferred solution and outcome.

Phase E. Sensitivity analysis and convergence

For analysing the sensitivity of an efficient solution to changes in the proportions of outcomes, a *multidimensional scan* of efficient solutions is implemented in some systems of DIDAS family. This operation consists in selecting an efficient outcome, accepting it as a base $\bar{q}^{bas}$ for reference points, and performing p'' additional optimization runs with the reference points determined by:

$$\bar{q}_j = \bar{q}_j^{bas} + \beta(\hat{q}_j^{uto} - \hat{q}_j^{nad}), \tag{31}$$

$$\bar{q}_i = \bar{q}_i^{bas}, \qquad i \neq j, \qquad 1 \leq j \leq p'', \tag{32}$$

where β is a coefficient determined by the user, $-1 \leq \beta \leq 1$. If the relative scaling is used and the reference components determined by (31) are outside the range $\hat{q}_j^{nad}$, $\hat{q}_j^{uto}$, they are projected automatically on this range. The reference components for stabilized outcomes are not perturbed in this operation. The efficient outcomes resulting from the maximization of the achievement function $s(q, \bar{q})$ with such perturbed reference points are typically also perturbed mostly along their subsequent components, although other their components might also change.

For analysing the sensitivity of an efficient solution when moving along a direction in the outcome space a *directional scan* of efficient outcomes can be implemented in systems of DIDAS family. This operation consists again in selecting an efficient outcome, accepting it as a base $\bar{q}^{bas}$ for reference points, selecting another reference point $\bar{q}$, and performing a user-specified number K of additional optimizations with reference points determined by:

$$\bar{q}(k) = \bar{q}^{bas} + \frac{k}{K}(\bar{q} - \bar{q}^{bas}), \qquad 1 \leq k \leq K \tag{33}$$

The efficient solutions $\hat{q}(k)$ obtained through maximizing the achievement function $s(q, \bar{q}(k))$ with such reference points constitute a cut through the efficient set $\hat{Q}_0$ when moving approximately in the direction $\bar{q} - \bar{q}^{bas}$. If the user selects one of these efficient solutions, accepts as a new $\bar{q}^{bas}$ and performs next directional scans along some new directions of improvement, he can converge eventually to his most preferred solution (Korhonen and Laakso, 1986).

4   Architecture of systems of DIDAS family

Several attempts have been made to define a Decision Support Systemas a *software system*. However, the situation is similar like in the case of defining the Decision Support Systemitself: very general definitions are too broad to be constructive. Following Ariav and Ginzberg (1985) Decision Support System must consists of modules providing the following functions:

- management of the dialogue between the user and the system,

- management of data,

- management of models.

Evidently, this definition is to broad to be constructive: for example, every statistical modeling package or modern simulation language supports all three functions. Also such software systems like spreadsheet or expert systems satisfy also this definition. Therefore, the above definition must be considered as a necessary condition – as the minimal functionality which Decision Support System must provide.

Decision Support Systems of DIDAS family belong to the class of Multiple Criteria Decision Support System. Therefore, this special aspect of the system will have impact on the architecture of such a system.

The gerneral structure of aspiration based Decision Support Systemsfollows from the presented above idea of aplying the achievement function to characterize and compute satisficing solution. The following steps must be performed to obtain response to aspiration specified by decision maker:

- The decision maker specifies aspirations for objectives,

- Using the specified aspirations and achievement function, the multiple criteria optimization problem is reduced to a single objective minimization problem,

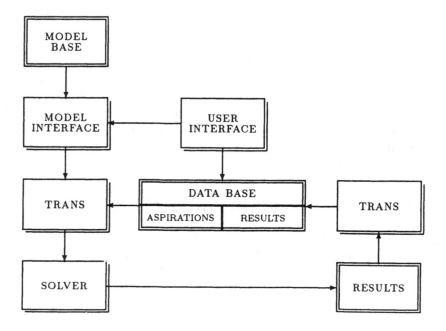

Figure 1: Structure of the aspiration–based Decision Support System

- The resulting single objective minimization problem is solved using one of the standard mathematical programming techniques,

- The obtained solution is converted to its original, multiple objective formulation, providing to the user information about objectives, attainability of aspirations etc.

Therefore, beside the standard modules specified by Ariav and Ginsberg (1985) as essential components of every Decision Support System, the aspiration based Decision Support Systemmust contain several specific modules (Figure 1)

- The *solver* which is the mathematical programming package capable of solving nonlinear programming problems resulting as the transformation of original multiple criteria formulation,

- The *transformer* which converts the multiple criteria problem to the single criterion one and reverse,

- The interface coordinating actions of the solver, transformers and the user–oriented part of the system.

Graphics in Decision Support Systems

Although Decision Support Systems are understood as interactive systems, not too much attention has been paid towards analysis of the role of graphics in building man–machine interfaces, designing architecture of these interfaces and principles of their operation. These issues are especially important due to availability of efficient computer hardware, modern computational algorithms and user friendly interface which allow (and stimulate) to use a Decision Support Systems as a generator of large amount of information. The amount of information is usually too big to allow full insight into all quantitative aspects of the decision problem being solved. Therefore, the basic idea of Decision Support Systems as a device supporting feedback between decision maker and formalized model is frequently not sufficiently explored.

There are two different aspects of graphical information presentation in Decision Support Systems:

- Presentation of large amount of data in graphic form to improve understanding of data, their internal structure as well as relationships between components of the decision problem being solved (objectives, decision variables etc.),

- Providing support for dynamic interaction between user and Decision Support Systems. This function contributes to the learning process by allowing analysis of the history of interaction process which has resulted in achieving the current state of the decision making process. The other function of the graphic interface is simplifying planning consecutive steps in elaborating a decision with help of Decision Support System.

Most of research in the field of graphic perception and graphic interaction design has been directed towards first of the mentioned above aspects of information presentation. Several guidelines for good graphic presentation of data have been proposed (for definition of *graphic excellence* see Tufte, 1983). Experimental study of graph perception has been performed by several researchers. Cleveland and McGill (1984) investigated the *graphical perception* as the *ability of visual decoding of information encoded on graphs.* On the basis of several experiments they have determined efficiency of various forms of graphic presentation as well as suggested possible ways of improvement of graphic presentation style.

Lucas (1981) analyzed the impact of computer–based graphics on efficiency of decision making. The experimental task consisted of selecting quarterly reorder quantities for an importer under condition of uncertain demand, using both graphic and tabular data presentation. Results of this research do not give precise answer regarding comparison of tabular and graphical data presentation. According to the author, in *analytic situation*

However, this way of computing the upper bound for efficient outcomes is not always practical due to required computational effort. Many systems of DIDAS family use a different way of estimating the utopia point. This way consists in subsequent maximizations of the achievement function $s(q, \bar{q})$ with suitably selected reference points $\bar{q}$. If an objective should be maximized and its maximal value must be estimated, then the corresponding component of the reference point should be very high, while the components of this point for all other maximized objectives should be very low. If an objective should be minimized and its minimal value must be estimated, the corresponding component of the reference point should be very low, while other components of this point are treated as in the previous case. If an objective should be stabilized and both its maximal and minimal values must be estimated, then the achievement function should be maximized twice, first time as if for a maximized objective and the second time as if for a minimized one (while the obtained maximal and minimal values will be denoted by $\hat{q}_i^{uto}$ and $\hat{q}_i^{nad}$, respectively, although it is difficult to say which of them corresponds to the concept of utopia point). Thus, the entire number of optimization runs in utopia point computations is $p'' + 2(p - p'')$. This is especially important in dynamic cases. It can be shown that this procedure gives a very good approximation of the utopia point $\hat{q}^{uto}$.

During all these computations, the lower bound for efficient outcomes can be also estimated. This can be done by recording the lowest efficient outcomes that occur in subsequent optimizations for maximized objectives and the highest ones for minimized objectives. However, such a procedure results in the accurate, tight lower bound for efficient outcomes – called *nadir point* $\hat{q}^{nad}$ – only if $p'' = 2$. For larger numbers of maximized and minimized objectives, this procedure can give misleading results. The accurate computation of the nadir point becomes a very cumbersome computational task requirinng very special approaches (see Isermann and Steuer, 1987).

Therefore, some systems of DIDAS family accept user-supplied estimates of lower bounds for objectives. They also offer an option of improving the estimation of the nadir point in such cases. This option consists in additional p'' maximization runs for achievement function $s(q, \bar{q})$ with the following rules for selecting the reference point if the objective in question should be maximized:

- reference point corresponding to the selected objective $\bar{q}$ is very low,

- reference points are very high for other maximized objectives and very low for other minimized objectives,

- stabilized objectives should be considered as floating.

If the objective in question should be minimized, the similar rules applies: the corresponding reference component should be very high, while other reference components should be treated as in the previous case. By recording the lowest efficient outcomes that occur in subsequent optimizations for maximized objectives and the highest ones for minimized objectives, a better estimation $\hat{q}^{nad}$ of the nadir point is obtained.

Interactive graphics for MCDM

According to the principle of exploratory data analysis the graphical user interface should allow performing a quick analysis of properties of data obtained in the process of solving decision problem with support of DSS. The situation is, however, different than in statistics – in DSS context the user can deal with historical data, but can also perform an *active experiment* running the DSS software to obtain new data to validate or invalidate his hypothesis. Therefore, the graphical data analysis interface should also provide access to all features of DSS.

However most of existing DSSs are equipped in modules for graphical data presentation, not too many of them are oriented towards exploratory data analysis.

The VISA approach proposed by Belton (1988) utilizes the concept of visual interactive modeling. According to Bell (see Belton, 1988 for further references) the Visual Interactive Modeling is the *process of building and using a visual interactive model to investigate issues important for decision maker.* The Visual Interactive Model has three essential components: a mathematical model, a visual display of the status of the model and an interaction device that permits the status of the model to be changed. The basic idea of VISA is the animation which allows displaying of changes of solution simultaneously with changes of certain parameters of the decision problem performed by the user (weighting factors).

The TRIMAP approach developed by Climacao (Climacao at all., 1988) has been designed for solving problems with 3 criteria. In each iteration two graphs are presented to the decision maker: the first one shows regions in weight space corresponding to each of the already known Pareto optimal vertices, the second one presents projections of Pareto solutions on a plane.

Korhonen and Laakso proposed extension of the *reference point* optimization to *reference direction* optimization (Korhonen and Laakso, 1986a, 1986b). Similar approach utilizing linear parametric programming has been proposed by Lewandowski and Grauer (Lewandowski and Grauer, 1982). The basic idea of this method is as follows:

- Decision maker specifies the reference direction, starting from the most recently obtained solution,

- The reference direction vector is projected on the Pareto surface and a curve traversing across the efficient frontier is obtained,

- Values of objectives along the projected reference vector are plotted on the screen using a distinct color or line pattern for each objective. The graphic cursor can be moved to any point on the curve and the corresponding numerical values of objectives are displayed.

In this way the decision maker can obtain an overview of the behavior of the objectives across the efficient frontier.

Further improvements of this concept have been implemented in the Pareto Race procedure (Korhonen, 1987, Kananen at all., 1990). Pareto Race is a dynamic version of the mentioned above method and allows interactive exploration of the Pareto surface.

However all the mentioned above methods provide various forms of dynamic interaction with Decision Support System, they do not support exploratory data analysis. These methods allow to use the interactive graphic interface to formulate questions about the problem being solved and to enter these questions to a computer to obtain answer, but they do not allow to investigate the *process of interaction*, since historical data are not stored and tools for their analysis are not available. Therefore, important questions regarding conflicts in objectives, their dependency and redundancy cannot be supported.

Reduction of dimensionality

The basic difficulty in visualization of results of MCDM analysis is created by the fact that dimensionality of the objective space is usually higher than 2. Therefore, several attempts have been made to perform 2 dimensional presentation of data located in higher dimensional space. This is not the goal of this paper to discuss all possible methods of such presentation. It is necessary, however, to mention two possible approaches:

- The *iconic approach*, when every point in n dimensional space is represented as an *icon* parametrized by values of coordinates of this data point. This technique has been adapted by Korhonen (1988) in his Harmonious Houses approach and implemented in VICO (Visual Multiple Criteria Comparison) system,

- The *projection approach*, when every point in n dimensional space is projected on a plane. The projection operator is selected in such a way that all important properties of data in n dimensional space (distance, correlation) are preserved for points projected on a plane with as high accuracy as possible. This technique is known as *Multidimensional Scaling* (see Kruskal and Wish, 1989) and has been adopted for MCDM by Marechal and Brans (1988) as a non–interactive data analysis procedure.

The iconic approach possesses several disadvantages. The iconic representation is not unique, since the same data can be transformed to an icon in many possible ways. Moreover, the analyst can "like" or "dislike" some shapes, even if these shapes are purely abstract and have no well defined meaning. In the case of houses or faces the bias can be even higher. Moreover, the only analytic technique for deriving conclusions about data being analyzed is a highly subjective visual inspection which cannot bring information about important quantitative characteristics of data like correlation.

However the projection technique is not free of problems, it can provide much more deep insight into structure of data. Among many existing projection techniques the *biplot* seems to be the best tool which ensures preservation of many characteristics of data important for decision making.

Principal component biplot

The *biplot* has been proposed by Gabriel (1971) as a convenient technique for graphical presentation of matrices of rank 2. Such a matrix Y of dimensions $m \times n$ can be factorized as a product of two matrices G and H

$$Y = GH' \tag{34}$$

where G is $n \times 2$ matrix and H is $m \times 2$ matrix. From the above formula follows that

$$y_{ij} = g_i' h_j \tag{35}$$

Therefore, this factorization assigns vectors $g_1, ..., g_n$ to rows of Y and vectors $g_1, ..., g_n$ to columns of Y. Since the vectors g and h are vectors of order two, they can be plotted on a plane giving a representation of the $m \times n$ elements of the matrix Y by means of the inner products of the corresponding vectors. Such a plot is called a *biplot* since it allows the rows and columns to be plotted jointly.

Let us consider the $n \times m$ matrix Y of data. In our case this matrix represents set of nondominated solutions of the MCDM problem with columns corresponding to objectives and rows corresponding to solutions.

It is well known, that such a matrix can be represented using the singular value decomposition

$$Y = \sum_{\alpha=1}^{r} \lambda_\alpha p_\alpha q_\alpha' \tag{36}$$

where λ_α denotes the singular values of the matrix Y, p_α denotes the singular column and q_α denotes the singular row satisfying the following relations:

$$p_\alpha' Y = \lambda_\alpha q_\alpha' \tag{37}$$

$$Y q_\alpha' = \lambda_\alpha p_\alpha' \tag{38}$$

$$YY' p_\alpha = \lambda_\alpha^2 p_\alpha \tag{39}$$

$$Y'Y q_\alpha = \lambda_\alpha^2 q_\alpha \tag{40}$$

It ia also known, that the matrix M of rank s

$$M_{(s)} = \sum_{\alpha=1}^{s} \lambda_\alpha p_\alpha q_\alpha' \tag{41}$$

can be considered as the least-squares optimal rank s approximation of matrix M.

Let us assume that the matrix Y is normalized, i.e. mean value of each objective has been subtracted from the corresponding column of the matrix Y, then

$$S = \frac{1}{n} Y'Y \tag{42}$$

can be interpreted as the variance-covariance matrix. Let us consider the rank 2 approximation of matrix Y

$$M_{(2)} \sim Y \tag{43}$$

and let $\{\lambda_1, \lambda_2\}$ denote two largest singular values. Than the matrix $M_{(2)}$ can be factorized as

$$M_{(2)} = GH' \tag{44}$$

where

$$G = (p_1, p_2)\sqrt{n} \tag{45}$$

$$H = \frac{1}{\sqrt{n}}(\lambda_1 q_1, \lambda_2 q_2) \tag{46}$$

Therefore, if we consider definition of the matrix $M_{(2)}$, the following conclusions can be derived from the general properties of the singular values decomposition and factorization

$$Y \sim GH' \tag{47}$$

$$YS^{-1}Y' \sim GG' \tag{48}$$

$$S \sim HH' \tag{49}$$

where $\sim$ stands for *is approximated by means of a least squares fit of rank two*. Therefore, we can establish the following approximation for values of matrix Y

$$y_{ij} \sim g_i' h_j \tag{50}$$

as well as for covariances, variances and correlation

$$s_{j,g} \sim h_j' h_g \tag{51}$$

$$s_j^2 \sim \parallel h_j \parallel^2 \tag{52}$$

$$r_{j,g} \sim \cos(h_j, h_g) \tag{53}$$

Moreover, the Euclidean distance between vectors $\{g_i\}$ approximates the standardized distance between observations

$$d_{i,j} \sim \parallel g_i - g_j \parallel \tag{54}$$

where

$$d_{i,j} = (y_i - y_j)' S^{-1}(y_i - y_j) \tag{55}$$

The expression

$$\frac{1}{n}\sum_{i=1}^{n}(y_{ij} - y_{ig})^2 \sim \parallel h_j - h_g \parallel^2 \tag{56}$$

gives an approximation to the average squared difference between variables.

Especially important for interpreting biplot are relationships (50) – (53). According to these formulas:

- elements of data matrix can be approximated as orthogonal projections of vectors h on principal component directions,

- correlation coefficients between variables are approximated by angle between vectors h,

- variance of the variable is approximated by length of corresponding vector h.

If solutions collected during interaction with the DSS are considered as data points and objectives as variables, the biplot technique can be used to provide the following information:

- *Clusters.* Since distance between solutions (data points) on biplot plane approximate these in original space (55), clusters are easily visible and can be detected by visual inspection of the biplot. Therefore, it is possible to analyze similarities and dissimilarities between solutions.

- *Variance.* Length of vector h approximates variance of the objective. This value can be interpreted as *flexibility* or *sensitivity* of the objective.

- *Correlation.* The angle between vectors h represents correlation between objectives. Analysis of correlation can support deriving the following conclusions:

  - *Redundancy of objectives.* If some objectives are highly positively or negatively correlated, they can be replaced by one or by lower number of objectives,

  - *Conflicts between objectives.* It can happen that two objectives are positively correlated, but the third one is negatively correlated with first two. In such a case it is not possible to improve all these objectives simultaneously.

The BIPLOT interface implements the formulas presented above and provides tools for graphical visual interaction (Figure 2). The objectives are scaled according to the principles of DIDAS methodology, i.e. using the *utopia* and *nadir* points. Both h and g vectors are presented on the screen. The user can move a cursor (using mouse or keyboard); simultaneously a point representing cursor is projected on directions generated by h vectors giving values of objectives corresponding to this point. The center of coordinates can be located in a point corresponding to mean values of objectives or moved to a point corresponding to selected solution. In the later case it is possible to analyze improvements of objectives with respect to a selected solution.
Several other options are available:

- Comparison of two solutions,

- Selection and removing a group of points,

- Displaying vectors h and g separately.

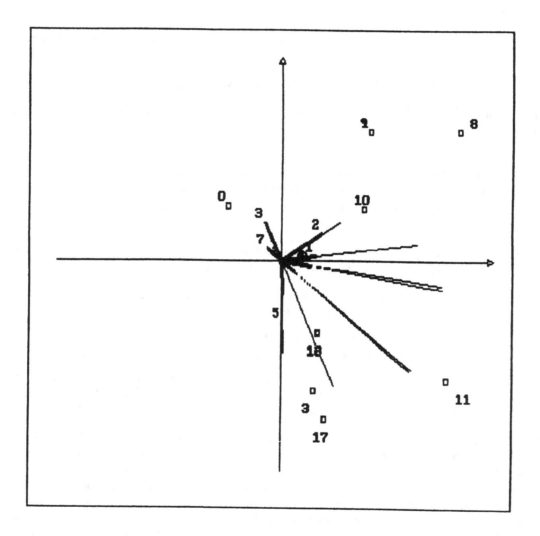

Figure 2: BIPLOT interface to aspiration-based DSS

IAC - DIDAS - L2 V5. Model editing			Bounds:		Names Units Value	Rolls 50 g	Cereals 50 g	Butter 10 g	Fruitfr 150 g
Names	Units	Value	lower	upper	upper lower	3.	2.	3.	2.
Cost				120.		6.	3.	5.	14.
Taste	artun		4.	40.		3.	1.	2.	4.
Stimulus	artun		4.	80.		3.	3.	4.	3.
Callorie	kcal		150.	1200.		124.	179.	75.	79.
Protein				60.		4.	3.	0.1	0.5
Carbohyd			10.	100.		26.	36.		11.
Fats				80.		1.	2.	8.	0.5
Calcium			180.	720.		8.	10.	2.	9.
Magnesiu				400.		12.	23.	0.2	5.
Phosphor				1000.		42.	103.	1.6	13.
Iron				16.		1.	1.		0.4
Vit.A			100.	1600.				270.	160.
Vit.B				3.		0 12	0.14		6.0e-2
Vit.C				200.					30.
Vit.PP				10.		0.4	1.	1.0e-2	0.23

Figure 3: Model definition in IAC–DIDAS–L system

Model management in DIDAS system

Specifying linear models

Table editor

In the principle, generation and manipulation of linear models is not complicated: a linear model can be represented as a set of linear equations and inequalities. These equations and inequalities can be represented as a matrix or a table. Therefore, in the simplest case a table manipulation tool (*table editor*) can be used for defining linear models. Such software tools for table manipulation are known as *spreadsheets* and are commercially available.

The general purpose spreadsheet can be conveniently used for generation a linear programming problem. Unfortunately, integration of a commercial spreadsheet with other software modules is difficult. Therefore, specialized tools for table manipulation needed for generation of linear programming problems have been developed.

Exactly such a tool is used in the IAC–DIDAS–L family of DIDAS systems. The user defines a model specifying rows and column names (names of constraints and names of variables, respectively) and enters numerical values in cells corresponding to non–zero elements (Figure 3).

The user can specify units and bounds for constraints and variables. All these data can be defined during editing phase and most of them cannot be modified in further phases. The only exception are bounds which can be overwritten in the interaction phase during definition of the multiobjective programming problem. This does not imply that bounds specified in the model definition are modified. In the fact, model bounds are just taken as starting values for bounds of the problem. Most interesting parts of the model definition can be examined in further phases of interaction with the system. This concerns names of rows and columns as well as units and bounds. Usually the layout of table editor is designed in such a way, that solution corresponding to active model can be displayed simultaneously with displaying model coefficients.

The table-oriented style of interaction has been used in many commercial microcomputer linear programming packages (Stadtler at al., 1988).

Unfortunately, simple table editor is not sufficient for solving more complex, real life problems. Evidently, this is always possible to define a linear programming model as a big table, but the following facts make such manual operation difficult:

- In real-life problems most of coefficients of the linear programming table are equal to zero. Therefore, when working with a big table a user has to scroll the screen very frequently to display the coefficient which he wants to define or update. Consequently, access to elements of linear programming table is not as easy as it could follow form the "user friendliness" of a table editor.

- Frequently, especially in the case of modeling dynamical systems, the linear programming table has a structure consisting of periodically repeated similar blocks corresponding to successive instances of time. This can be a very time consuming process to enter all elements of such blocks manually if table editor does not provide specialized tools for supporting such a task.

- Elements of linear programming table can depend on other data constituting a part of problem formulation. These relationships cannot be programmed using a simple table editor.

Therefore, the simple table editor implemented in IAC-DIDAS-L system can be used only for solving relatively small and simple problems. However, despite all limitations, the existing implementation of table editor provides relatively high level of flexibility of model management.

MPS format and model generators

Most commercial systems designed for solving linear programming problems follows the data format originally designed to the MPS series of linear programming systems developed for IBM computers. This format has become de facto standard adopted by developers of many other linear programming systems, including early versions of DIDAS system.

The data that correspond to a linear programming problem is grouped in the following sections:

- NAME This section consists of the keyword NAME and a field containing name of the problem

  name   mann02

- ROWS In this section all the rows names are defined together with the row type. The row type in entered together with symbolic name of this row. The following row types are admissible:

  - E equality
  - L less than or equal
  - G greater than or equal
  - N free (no restriction). The first free-type row encountered is regarded as the objective row, unless the objective is explicitly identified in the specification file or by a code itself. In DIDAS system every free row can be converted into the objective row.

```
rows

 e  kap...01

 e  kap...02

 l  mon...01

 l  mon...02

 g  cka...01

 g  cka...02

 n  goal
```

- COLUMNS This section defines the variables and the coefficients of the constraints matrix (including the objective row). Only nonzero coefficients are entered. The data are entered column by column and all data for nonzero entries in each column must be grouped together contiguously. The column entry contains the column name, the row name and the value of coefficient.

```
columns

  con...01  goal              0.95
```

```
con...01  mon...01           1.00

con...02  goal               0.90

con...02  mon...02           1.00

inv...01  kap...01           1.00

inv...01  mon...01           1.00
```

- RHS This section contains the non-zero elements of the right-hand sides of constraints. The data format corresponds to COLUMNS section, the only difference is due to replacement of a column name by a label (with may also be blank). More than one right-hand side set may be specified in this section (the one to be used for the current run is specified in the specification file by its label).

```
rhs

   test1    cka...01         3.16

   test1    cka...02         3.16

   test2    cka...01         4.11
```

- RANGES This section contains entries for inequalities rows for which both lower and upper bound exist. The data format is the same as for RHS section. The value of range is interpreted as the difference between the upper and the lower bound for the respective row. One of those bounds (if nonzero) is entered in the RHS section, the type of the row indicates whether lower or upper bound in defined in the RHS section.

- BOUNDS This section contains changes for bounds for variables initially set to default values. The default bounds are usually defined as 0 for the lower bound and no constraint for the upper bound. More than one set of bounds may be specified in this section (the one to be used for the current run is specified in the specification file by its label). More than one bound for a particular variable may be entered. The bound indicators have the following meaning:

  - LO lower bound

  - UP upper bound

- FX fixed value for the variable
- FR free variable (no bounds)
- MI no lower bound, upper bound equal to 0.
- PL no upper bound, lower bound equal to 0.

```
bounds

    up bnd1       inv...01          0.17

    up bnd1       inv...02          0.17

    lo bnd1       con...01          0.65
```

Although conceptually simple, the MPS format cannot be used for manual model definition. This is due to fixed format of every entry as well as due to the fact that even specification of simple models can contain several hundred if not thousand lined of MPS code. Therefore, this code can be used only as intermediate code. Therefore, special tools for generating MPS file are necessary.

In the simplest case, the person involved in modeling task (*the analyst*) writes program in general purpose programming language to convert abstract formulation of a model into the corresponding MPS file. This program, named *model generator* can read data from primary data file and incorporate these data into model description. This procedure allows easy experimenting with a model. Moreover, the MPS file produced by model generator can be further transformed to incorporate some other features into model description, like required by DIDAS methodology transformation of multiple criteria problem into a single criterion one. This approach has been applied in several versions of DIDAS system and successfully applied in solving practical problems (MESSAGE model, Messner, 1984).

The basic disadvantage of model generators is connected with complexity of programming such a device and the lack of flexibility of the received product. Every change of the model specification requires corresponding changes in model generator code what is usually complicated and time consuming task difficult to perform by anybody but the author of the program. Moreover, validation of the correctness of generated model description is difficult.

Modeling languages

To simplify the task of model specification, several software tools called *modeling languages* have been developed. According to Dolk (1983) the linear programming modeling language should fulfill the following requirements:

- It must be possible to express any linear programming problem in its initial, algebraic form by use of the modeling language,

- It must be possible to create a convert the algebraic description of the problem into internal or intermediate representation which can be accepted by linear programming solver.

We will not discuss here the state-of-the art of this discipline (see, for example Dolk, 1986). Instead, we will present briefly only one particular modeling language which can be conveniently used with HYBRID system and other version of DIDAS systems accepting the MPS format for describing linear programming problems (Huerlimann, 1987). This modeling language, named LPL is a modification of Pascal programming language, allows to define linear programming problems using a standard mathematical notation. Without making an attempt to fully present this language, we will give only a simple example of this notation and a corresponding MPS file. The notation is straightforward and easy to understand, even for nonspecialist (see Figure 4). The LPL compiler produces executable program which produces description of the problem in MPS format or other format accepted by commercial linear programming packages. The MPS file generated by LPL system is presented on Figure 5. The language possesses more advanced features, like indexed notation and set variables. A variable $Q_{p,t}$ is written as Q(p,t). If we define

SET

```
p = (p1:p3)

t = (1:4)

    ...SUM[p,t] Q[p,t]...
```

the resulting action performed by LPL program will generate a fragment of linear programming problem equivalent to the following formula

$$\sum_{p=1}^{3}\sum_{t=1}^{4} Q_{p,t}$$

Several other constructs are available in LPL which allow easy definition of linear programming models. In particular, this language provides easy access to database files created by commercial data base systems. Due to the character of output produced by LPL it can be used with systems accepting MPS format as problem specification language.

Specifying nonlinear models

The issue of designing a problem interface for nonlinear version of DIDAS is especially important and more difficult than in a case of linear programs. This is due to the fact that table representation is not sufficient – a model can be described by set of nonlinear equations. Moreover, to prepare model for submitting it to the solver several algebraic manipulation are necessary. These manipulations include also algebraic differentiation of model equation.

A DIDAS/N system developed by Grauer and Kaden (1984) was the first published nonlinear version of such a system. It was based on MINOS/Augmented (Murtagh and Saunders, 1980) nonlinear programming system, an extended version of linear MINOS.

The interface used in Grauer's and Kaden's version of the system is rather conceptually simple, but not very user-friendly. The equations describing objective functions and constraints must be programmed in FORTRAN. The authors supply the *skeleton* or *template* of FORTRAN subroutine with empty holes where the user must locate FORTRAN code describing his problem. This is a rather complicated task – separate parts of the problem definition must be located in various places of the code, and the code itself must be written taking into account names of variables and structure used in this skeleton subroutine. What makes defining the problem especially difficult is the fact that writing his code the user must properly augment all his formulas with penalty function terms and their derivatives. This is conceptually rather difficult for a user which is not familiar with mathematical programming algorithms and can lead to numerous errors. Conflicting variable names are also probable.

The general purpose version of nonlinear DIDAS developed by Kreglewski and others (Kreglewski, at al., 1985) also needs a FORTRAN subroutine containing the problem description. Unlike to the previous system, the user must preserve only the general structure of the subroutine header (declaration of formal parameters) and COMMON block. No variable conflict can occur, and the standards according to which the body of the subroutine must be composed are quite clear and straightforward. The main disadvantage of this interface relates to the definition of derivatives – the user must calculate these derivatives analytically. This is usually a time consuming process and the source of various errors that are difficult to detect.

To minimize the probability of occurrence of errors in analytical gradient computation, several authors proposed numerical procedures for gradient checking. This approach was utilized by Kreglewski (1985) in his version of DIDAS system.

The simplest interface has the DIDAS–like system for solving water management problems developed by Kaden and Kreglewski. This system was designed to solve only one class of problems, therefore a model of the system was programmed only once, in a very efficient way. The user interacts with the system only on the level of input data and reference point selection.

Concluding, the following are the basic disadvantages of the existing implementation of problem interface in nonlinear DIDAS:

- The user must compute analytically all derivatives of objective and constraint functions.

- The objective and constraint functions as well as all derivatives must be programmed in FORTRAN according to the specification supplied by the implementator of the system. This specification can be difficult to understand for nonexperienced user.

- The user must be familiar with every details of the computing environment which he is working with – such as program editor, compiler, linker, operating system command language etc. This is the most severe limitation, which restricts essentially the usability of the system. Usually, a long training is necessary to work with the computer efficiently and without troubles on this level of interaction.

- Any changes of the model – the process being in the fact one of the important stages of interactive work with the a Decision Support System, cannot be performed using tools provided by a Decision Support System.

It is necessary to point out that the model definition and modification is one of the most important stages of working with any decision support system. The sequence: program editor – FORTRAN compiler – linker – operating system, being in the fact one of the stages of interaction with the system, slows down essentially the interaction process, makes it difficult and inefficient. Therefore, the user friendliness of the problem interface requires special attention, especially in the case of nonlinear problems. This issue was the main design criterion of IAC–DIDAS–N system.

To simplify the process of defining nonlinear models, two approaches have been elaborated. The first one (Lewandowski, 1986) bases on the principle of extending and modifying the general purpose programming language like Pascal. In the fact, a small subset of Pascal has been considered for pilot implementation.

From the user's point of view, this version of Pascal constitutes a standard language with three additional declarations:

- **vardec** for declaring *decision variables,*

- **varcon** for declaring *constraints variables,*

- **varobj** for declaring *objective variables.*

These variables (named *functional variables*) can be treated like other, ordinary variables. They constitute the *interface* to the solver and other part of the system. The compiler, translating the program to the internal form, computes analytically all necessary derivatives, generates code for computing values of functions and stores this code. When the solver requests values of functions and derivatives, it sends values of arguments to *vardec* variables, and invokes the interpreter. The interpreter executes the code performing additional calculations necessary for obtaining values of derivatives and resulting values assigns to corresponding *varcon* and *varobj* variables.

To be more specific, let us consider the example in which SR, SI, TL and MRR are objective functions, and v, f and d – decision variables:

$$\ln SR = 7.49 - 0.44 \ln v + 1.16 \ln(1000f) - 0.061 \ln(1000d) \qquad (57)$$

$$\ln SI = -4.13 + 0.92 \ln v - 0.16 \ln(1000f) + 0.43 \ln(1000d) \tag{58}$$

$$\ln TL = 21.90 - 1.94 \ln v - 0.30 \ln(1000f) - 1.04 \ln(1000d) \tag{59}$$

$$\ln MRR = -11.33 + \ln v + \ln(1000f) + \ln(1000d) \tag{60}$$

The following constraints augment definition of the problem:

$$600 \le v \le 1200 \tag{61}$$

$$0.02 \le f \le 0.018 \tag{62}$$

$$0.05 \le d \le 0.10 \tag{63}$$

The corresponding program is presented on Figure 6.

The upper and lower limits for constraints variables are defined in the table-oriented part of the model management interface. The program editor, compiler and interpreter of extended Pascal are parts of the problem interface and from the user's point of view they constitute the integrated system.

The similar approach has been implemented in the IAC-DIDAS-N system, however the model management is not *language oriented* but *spreadsheet oriented*. Construction of the spreadsheet is similar to spreadsheets available commercially, however with special features following from the requirements of the Decision Support System. The user can define formulas according to the standard mathematical notation (Figure 7) with extensions including conditional statements. All necessary analytical transformations and numerical computations are performed by the system. In one particular variant of this system (IAC–DIDAS–G) the user can display all derivatives in analytical form.

Management of data

Except of functions supporting specifying of a model the problem interface should support *data management* including storage, manipulation and retrieval of all elements of decision problem being solved.

All data used by DIDAS system are divided in three groups according to interdependencies between variables as well as sequence of steps in interactive problem analysis:

- *Model* defined in *model edition phase* consisting of data defining components of the decision situation being supported: names, units and bounds for all input and output variables together with coefficients of the linear mathematical model or equations of a nonlinear model.

- *Problem* defined in *interaction phase* containing status of each outcome variable that defines its type, as well utopia and nadir points calculated for this combination of objectives. In multiobjective analysis each output variable can be used as objective function and therefore *minimized, maximized* or *stabilized* or as simple constraint (marked as *floating* or with undefined status). Alternative definition (*floating*) in the status field for variables acting as constraints is used to enable displaying them on the bar chart.

- *Result* consisting of reference point, scaling variables, solutions in objective space and in decision space computed by the system.

All the mentioned above components of the problem are accessible for the user. During interactive analysis many problems corresponding to one model can be generated as well as many results can be generated corresponding to a given problem. Some of them are significant and should be saved for further analysis. To make data generation and retrieval more efficient and less troubling for the user, the following conventions have been established:

- When model definition phase is completed, the model should be *locked* to ensure that all further experiments will be performed in consistent way, i.e. with the same model. This is important for consistency of results analysis and comparisons between various problems and results related to the same model.

- A new *problem* is generated every time the user changes the status of any outcome variable.

- A new *result* is created each time the reference point or scaling coefficients are modified and corresponding solution computed by the system.

All the mentioned above elements can be stored on the disk as DOS files. Every group (i.e. model, problem and result) is stored in one file. To avoid name conflicts special naming convention has been established. Every file keeps information about related files. For example, file containing result stores also name of a file containing corresponding problem. Such links make possible manipulation of all related files without user's intervention and ensure loading data in correct order. Every file stores also information creation time of the item (model, problem or result) to avoid loading old result files with a new problem definition.

The similar approach has been applied in other implementations of decision support systems of DIDAS family.

5 Implementation of decision support systems of DIDAS family

Linear systems

System MOCRIT

An early, prototype linear version was developed by Kallio, Lewandowski and Orchard-Hays (1980). This version utilized the professional LP package SESAME and has been implemented using the specialized programming language DATAMAT. The system consists of the series of modules converting the multiple criteria problem into the equivalent single criterion one and back. Since the SESAME/DATAMAT is available only on the IBM–370 mainframe computers, the MOCRIT system was not transferable. The user interface was rather poor and the usage of the system was limited to its authors and their collaborators.

System LPMULTI

The second, also linear, version of DIDAS family systems was developed by Lewandowski (1982). It was designed as pre- and postprocessor programs to a commercial LP package with standard MPSX input and output. This system consists of three modules:

- the interactive *editor* for manipulation the reference point and the objectives,

- the *preprocessor* which converts the input model file prepared in standard MPSX format, containing the model description into its single criterion equivalent,

- the *postprocessor*, which extracts the information from the LP system output file, computes the values of objectives and displays the necessary information.

The transformed problem (single criterion) is formulated by the system using the standard notation (MPSX). Therefore, any commercial Linear Programming package can be used for solving such a problem. Due to such design, the system was easily transferable and many practical problems were solved using it on various computers (Grauer and Brillet, 1982). The main drawback of this implementation was that the interface between pre- and postprocessor and a the LP solver was based on reading and writing disk files. This process can be time consuming for larger problems. An interaction with the user was very simple but inconvenient because of long time responses of the system transferring large amount of data.

This system, available in version for VAX-UNIX and for MS-DOS has been distributed to several scientific centers in the world and used for solving practical problems.

System MM–MINOS

The design goal of the next version of DIDAS was to eliminate, if possible, disk transfers and changes of data structures inside the system. It was done by Kreglewski and Lewandowski (1983) as a interactive multicriteria extension of MINOS linear programming system (Murtagh and Saunders, 1977). The reference point concepts were implemented accessing MINOS internal data structures. The user interface was redesigned and many new options added. However, the portability problems has been not solved since MINOS is not easily transferable.

System DIDAS–MZ

The experiences of existing developments as well as portability problem led to development of a new versions of DIDAS called DIDAS-MZ. DIDAS-MZ is based on a linear programming solver from IMSL library which is widely accessible. Moreover, since the system has been implemented is standardized FORTRAN–77 any other optimization routine can be easily integrated with this system. Therefore, DIDAS-MZ is much easier transferable than previous versions (Lewandowski at al., 1985).

System IAC–DIDAS–L1

In 1986, a new generation of DIDAS family systems was initiated, designed for work on IBM-PC-XT and compatible computers. These are: IAC-DIDAS-L1, IAC-DIDAS-N as well as other systems. More detailed descriptions can be found in the paper by Lewandowski (1988) as well as in the paper by Lewandowski and Wierzbicki (1988).

The representative of this generation of systems of DIDAS family is the IAC–DIDAS–L1 system. This version of the system is the decision analysis and support software for interactive, multiobjective analysis for *models describing substantive aspects of a decision situation* that are represented *in linear programming or dynamic linear programming form*.

IAC–DIDAS–L1 system *can be used either by analysts* specialized in modelling *or by decision makers* experienced in a given field but not necessarily computer specialists. This system help in organizing work with decision models in a process of interactive, dynamic decision support, providing tools for performing the following steps:

- model edition and initial analysis,

- formulation of a multiobjective decision analysis problem,

- initial assessment of bounds of decision outcomes or objectives for a given problem.

The IAC–DIDAS–L1 system consists of three functional modules:

- *The Problem Interface* used for specification of the model of the decision problem. Since the model is formulated in terms of linear programming language, a simple spreadsheet is used as such interface. A model of multiobjective linear programming type is characterized by its decision variables, its outcome variables defined by linear model equations, and its constraints or bounds on various variables. In multiobjective analysis, *the user can select objective variables* that might be *minimized, maximized* or *stabilized* close to given values.

  The defined model can is treated by the system as a *virtual* one – several details of the model formulation can be left unspecified. When such details are specified (e.g. bounds for variables), the *instance of the problem* is being generated. Therefore, several *problems* being instantiations of a given model can simultaneously exist in the system.

- *The Interaction Interface* used for interactive analysis of the *problem*. During such analysis several experiments with a problem can be performed by changing aspiration levels or some other parameters of the problem. The system keeps a track of consecutive results of analysis and stores results marked by the user as important in a *result directory*. Because the user can formulate various multiobjective decision problems for given model (by maximizing or minimizing multiobjective various model outcomes, etc.), the system keeps also track of various problem formulations. Finally, the system allows also to work with various models of one or more decision situations. Therefore, the system has also the *model and problem directory*.

  Since for every problem a series of *results* can be generated, the relationships between various information generated by the system have hierarchical structure.

- *The Information and Data Management Interface* used for manipulating all information generated by the system and specified by the user. This includes manipulation of the Model Base, the *Problem Base* and the *Result BAse* which provide access to all elements of the decision problem, both created by the user and generated by the system.

An interactive multiobjective analysis of the problem based on the principle of reference point optimization is performed. The user (the analyst or the decision maker) indicates the type of solutions that he is interested in by specifying his aspiration levels for objective outcomes and the decision support system responds to his directions by solving optimization problem and answering, whether his aspirations are attainable. If not, the system proposes decisions with outcomes that come uniformly as close as possible to the stated aspiration levels. If the aspiration levels are attainable but cannot be exceeded, the system proposes decisions with outcomes that precisely match the aspiration levels. If the aspiration levels can be exceeded, the system proposes decisions with outcomes that uniformly exceed the aspirations. By changing the aspiration levels, *the user can easily control and select such decision options that are best suited for his preferences.*

System DINAS

DINAS (Ogryczak at al., 1988) is a decision support system which enables the solution of various multiobjective transshipment problems with facility location. The distribution-location type problems belong to the class of most significant real–life decision problems based on mathematical programming. They are usually formalized as the so–called transshipment problems with facility location.

A network model of the transshipment problem with facility location consists of nodes connected by a set of direct flow arcs. The set of nodes is partitioned into two subsets: the set of fixed nodes and the set of potential nodes. The fixed nodes represent "fixed points" of the transportation network, i.e., points which cannot be changed whereas the potential nodes are introduced to represent possible locations of new points in the network. Some groups of the potential nodes represent different versions of the same facility to be located (e.g., different sizes of a warehouse etc.). For this reason, potential nodes are organized in the so–called selections, i.e., sets of nodes with the multiple choice requirement. Each selection is defined by the list of included potential nodes as well as by a lower and upper number of nodes which have to be selected (located).

A homogeneous good is distributed along the arcs among the nodes. Each fixed node is characterized by two quantities: supply and demand on the good, but for mathematical statement of the problem only the difference supply–demand (the so–called balance) is used. Each potential node is characterized by a capacity which bounds maximal flow of the good through the node. The capacities are also given for all the arcs but not for the fixed nodes.

A few linear objective functions are considered in the problem. The objective functions are introduced into the model by given coefficients associated with several arcs and potential nodes. They will be called cost coefficients independently of their real character. The cost coefficients for potential nodes are, however, understood in a different way than for arcs. The cost coefficient connected to an arc is treated as the unit cost of the flow along the arc whereas the cost coefficient connected to a potential node is considered as the fixed cost associated with activity (locating) of the node rather than as the unit cost.

Summarizing, the following groups of input data define the transshipment problem under consideration:

- objectives,

- fixed nodes with their balances,

- potential nodes with their capacities and (fixed) cost coefficients,

- selections with their lower and upper limits on number of active potential nodes,

- arcs with their capacities and cost coefficients.

In the DINAS system we placed two restrictions on the network structure:

- there is no arc which directly connects two potential nodes;

- each potential node belongs to at most two selections.

The first restriction does not imply any loss of generality since each of two potential nodes can be separated by an artificial fixed node, if necessary. The second requirement is not very strong since in practical models usually there are no potential nodes belonging to more than two selections.

The problem is to determine the number and locations of active potential nodes and to find the good flows (along arcs) so as to satisfy the balance and capacity restrictions and, simultaneously, optimize the given objective functions.

For handling multiple objectives DINAS utilizes an extension of the reference point approach. A special TRANSLOC solver has been prepared to provide the multiobjective analysis procedure with solutions to single–objective problems. The solver is hidden from the user but it is the most important part of the DINAS system. It is a numerical kernel of the system which generates efficient solutions. The concept of TRANSLOC is based on the branch and bound scheme with a pioneering implementation of the simplex special ordered network algorithm with implicit representation of the simple and variable upper bounds.

DINAS is equipped with the built–in network editor EDINET. It is a full–screen editor specifically designed for input and edit data of the network model of the transshipment problems with facility location. The essence of the EDINET concept is a dynamic movement from some current node to its neighboring nodes, and vice versa, according to the network structure. The input data are inserted using window-oriented editor while visiting nodes.

System HYBRID

HYBRID (Makowski and Sosnowski, 1988) is a member of DIDAS family of decision analysis and support systems which is designed to support multicriteria analysis via reference point optimization. HYBRID can be used by an analyst or by a team composed of a decision maker and an analyst or – on last stage of application – by a decision maker alone. HYBRID is a tool which helps to choose a decision in a complex situation in which many options may and should be examined.

HYBRID 3.1. includes all the functions necessary for the solution of linear programming problems, both static and dynamic (in fact also for problems with structure more general then the classical formulation of dynamic linear problems). HYBRID 3.1. may be used for both single- and multi-criteria problems. The package may be also used for solving single-criteria linear-quadratic problems. Since HYBRID is designed for real-life problems, it offers many options useful for diagnostics and verification of a model for a problem being solved.

HYBRID uses a non-simplex algorithm for solving linear programming problems. The LP problem is solved by minimizing a sequence of quadratic functions subject to simple constraints (lower and upper bounds). This minimization is achieved by the use of a method which combines the conjugate gradient method and an active constraints strategy. The method exploits the sparseness of the matrix structure. A dynamic problem is solved through the use of adjoint equations and by reduction of gradients to control subspaces. The simple constraints (lower and upper bounds for non-slack variables) for control variables are not violated during optimization and the resulting sequence of multipliers is feasible for the dual problem. Constraints other then those defined as simple constraints may be violated, however, and therefore the algorithm can be started from any point that satisfies the simple constraints.

The HYBRID system offers the following features:

- Input of data and the formulation of an LP problem follow the MPS standard,

- Modification of the problem at any stage of its solution by changing the matrix of coefficients, introducing or altering right-hand sides, ranges or bounds,

- Formulating and solving the multicriteria problem as a sequence of parametric optimization problems modified in interactive way upon analysis of previous results,

- Selection of the solution technique suitable for static or dynamic problems as well as special scaling algorithms for badly conditioned problems,

- Comprehensive diagnostics of correctness of model formulation as well as tools for model correction, update and modification.

The package is constructed in modular way to provide a reasonably high level of flexibility and efficiency. This is crucial for a rational use of computer resources and for planned extensions of the package and possible modification of the algorithm. The package consists of four modules:

- The *preprocessors* that serve to process data, enable a modification of the model, perform diagnostics and may supply information useful for the verification of a model,

- The *optimization module* for solving a relevant optimization problem (either static or dynamic),

- The *postprocessor* that provides results in the standard MPS format and generates the file which contains all information needed for the analysis of a solution,

Moreover, the PC version of HYBRID 3.1 contains additionally a driver which eases the usage of all subpackages. The driver provides a context sensitive help which helps an inexperienced user in efficient usage of the package.

Nonlinear systems

DIDAS–N system

A DIDAS/N system developed by Grauer and Kaden (1984) was the first published nonlinear version of such a system. It was based on MINOS/Augmented (Murtagh and Saunders, 1980) nonlinear programming system, an extended version of linear MINOS. Unfortunately, this solver is not robust and efficient enough for realistic nonlinear programming problems. Moreover, the user interface in the DIDAS/N system was rather complicated, hence applications of this system were rather limited. Earlier versions of DIDAS were also adapted for special purposes by Strubegger and Messner (Strubegger, 1985, Messner, 1985).

DIDAS–NL system

Lewandowski and Kreglewski (1985) developed another, general purpose nonlinear version of DIDAS system. It was based on a solver from Modular System for Nonlinear Programming (Kreglewski et al., 1984) and written completely in FORTRAN, hence easily transferable to arbitrary computer. The user interface was reasonably simple, but preparation of data for the system was not quite straightforward. Since the system utilizes its own solver and has been implemented is a standardized FORTRAN-77 it possess high level of portability and has been successfully implemented in several scientific centers.

IAC–DIDAS–N

The IAC–DIDAS–N (Kreglewski at al., 1988) package is designed to support models of multiobjective nonlinear programming type. Models of this type include two classes of variables:

- *input variables* that can be subdivided into *decision variables* (means of multiobjective optimization) and *parametric variables* (model parameters that are kept constant during multiobjective analysis but might be changed during parametric or sensitivity analysis),

- *outcome variables* that can be subdivided into several types, the most important of them being *optimized outcomes* or *objectives* (that can be either maximized or minimized or stabilized, that is, kept close to a desired level).

The user might change the classification of outcome variables and select his objectives among various outcome variables when defining an multiobjective analysis problem.

For all input and outcome variables, a reasonably defined nonlinear model should include:

- *lower* and *upper bounds* of these variables, including these variables representing objectives,

- *model equations* that define the dependence of all outcome variables on input variables. To make the model definition easier for the user, it is assumed that outcome variables are defined consecutively and that they can depend not only on input variables, but also on previously defined outcome variables.

The IAC–DIDAS–N system helps in definition, edition, initial analysis and verification, optimization and multiobjective decision analysis of a broad class of nonlinear models. An important feature of IAC–DIDAS–N is that it supports also automatic calculations of all derivatives of nonlinear model functions.

In the first phase, a user – typically, an analyst – defines the substantive model and edits it on the computer. IAC–DIDAS–N supports the definition and edition of substantive models in an easy but flexible standard format of a spreadsheet, where the input variables correspond to spreadsheet columns and the outcome variables – to spreadsheet rows. Another new feature of IAC–DIDAS–N is a *symbolic differentiation facility* that supports automatic calculations of all derivatives required by a nonlinear programming algorithm. The user does not need to laboriously calculate many derivatives and to check whether he did not make any mistakes. He defines model equations or outcome functions in a form that is acceptable for the symbolic differentiation program which admits functions from a rather wide class. The spreadsheet format allows also for display of computed values of automatically determined formulae for derivatives in appropriate cells.

The user of IAC–DIDAS–N can also have several substantive models recorded in special model directories, use old models to speed up the definition of a new model, etc., while the system supports automatically the recording of all new or modified models in an appropriate directory. Therefore, information manipulated by this system has a similar structure like in the case of IAC–DIDAS–L1 system.

In further phases of work with DIDAS-type systems, the user specifies a multiobjective analysis problem related to his substantive model and participates in an initial analysis of this problem. There might be many multiobjective analysis problems related to the same substantive model: the specification of a multiobjective problem consists in designating types of model outcomes, especially objective outcomes that shall be optimized, and specifying bounds on outcomes.

One of main functions of DIDAS-type systems is to compute efficient decisions and outcomes following interactively various instructions of the user and to present them to the user for analysis. This is done by using the principle of reference point optimization. Following the experiences with previous versions of nonlinear DIDAS systems, a special robust nonlinear programming algorithm was developed for IAC–DIDAS–N.

IAC–DIDAS–N utilizes the aspiration and the reservation levels as parameters in a special achievement function coded in the system. Its solver is used to compute the solution of a nonlinear programming problem equivalent to maximizing this achievement function, and responds to the user with an attainable, efficient solution and outcomes that strictly correspond to the user-specified references.

6   Applications of systems of DIDAS family

The first implementation of systems of DIDAS family was devoted to the application in forecasting and planning of the development of Finish forestry and forest industry sectors, based on a linear dynamic model of interaction between these two sectors (Kallio et al., 1980). Later, another version of DIDAS systems was applied (Grauer et al., 1982) to planning of energy supply strategies, which led to other applications in the analysis of future energy – economy relations in Austria (Strubegger, 1985) and of future gas trade in Europe (Messner, 1985).

Parallely, other research has been initiated including applications to forecasting and planning agricultural production in Poland (Makowski and Sosnowski, 1984), regional investment allocation in Hungary (Majchrzak, 1982), chemical industry planning (Gorecki et al., 1983). A specialized version of linear dynamic DIDAS was adapted to flood control problems (Lewandowski et al., 1984b). A nonlinear version of DIDAS was first applied to issues of macroeconomic planning (Grauer and Zalai, 1982) and later to problems of environmental protection of ground water quality (Kaden and Kreglewski, 1986).

It is rather not possible to discuss here all these applications. Instead, we will provide a detailed description of the system for supporting decision processes in control of flood distribution in a river network.

Flood control problem

This application has been performed in cooperation with the Institute of Automatic Control, Warsaw University of Technology (Lewandowski at al., 1985). The system under study is a river network consisting of the main river and three reservoirs (Figure 8). This is a simplified subsystem of the upper Vistula river in Poland. As it was mentioned above, the system consist of 3 general–purpose reservoirs supplying water to the main river. The goal of the system dispatcher during a flood period is to release water from reservoirs in such a way that peaks of the flood wave on the main river are as low as possible. Simultaneously, discharge from the reservoir should as low as possible to avoid damages downstream of the reservoir. In the fact, the control action delays or speeds up flood waves coming from reservoirs in a way which provides maximum level of desynhronization of these waves in main river.

It would be possible to formulate the above control or dispatching task as a single–criterion optimization problem. However, to provide such a formulation a performance function must be available. Unfortunately, function describing potential losses caused by a flood is extremely difficult for formulation and identification. Therefore, even if approximation of such function is provided, the resulting decision computed by a decision support system usually cannot be accepted by experienced system dispatcher.

The source of difficulties is connected with the fact that parameters of the flood wave are not the only quality factors which must be taken into account by a decision maker. A number of other factors, some of which are very difficult to formalize also play essential role in the decision process. From this reason, the system dispatchers formulates his goals rather in terms of a *desired shape of trajectory* than as minimization or maximization of a single performance index. Therefore, the DIDAS methodology seems to be the adequate tool for providing decision support for operating such a system.

The basic difficulty in applying DIDAS methodology for solving the flood control problem is caused by the complexity of the model describing propagation of a flood wave in a river network.

Mathematical model

According to the results of research regarding flood wave propagation in a channel, this process can be described with satisfactory accuracy by the following linear partial differential equation:

$$\frac{\partial p_i}{\partial t} = -C_i(x)\,\frac{\partial Q_i}{\partial x} \tag{64}$$

$$\frac{\partial p_i}{\partial x} = -D_i(x)\,Q_i \tag{65}$$

$$x \in [0,\ L_i] \tag{66}$$

In the above equation Q_i denotes the water flow in the i-th river segment and p_i denotes the water level. Clearly, both functions p_i and Q_i depend on space variable x and time variable t as their arguments, i.e.

$$p_i \equiv p_i(x,t) \tag{67}$$

$$Q_i \equiv Q_i(x,t) \tag{68}$$

Functions $C_i(x)$ and $D_i(x)$ depend on geometry of the river bed. The above equation can be easily transformed to the following form

$$\frac{\partial p_i}{\partial t} + \xi_i(x)\,\frac{\partial p_i}{\partial x} = \eta_i(x)\,\frac{\partial^2 p_i}{\partial x^2} \tag{69}$$

Appropriate boundary and initial conditions must augment equations describing every section of the river. Moreover, equation describing water balance in a node must be formulated to obtain a complete mathematical description of the system. If we consider a simple node with structure like on Figure 9, the following conditions must be satisfied:

$$p_1(t) = p_2(t) = p_3(t) \tag{70}$$

and

$$Q_1(t) + Q_2(t) = Q_3(t) \tag{71}$$

Therefore, we can formulate the following boundary conditions for this node

$$p_1(L_1,t) = p_3(0,t) \tag{72}$$

$$p_2(L_2,t) = p_3(0,t) \tag{73}$$

$$Q_3(0,t) = Q_1(L_1,t) + Q_2(L_2,t) \tag{74}$$

This set of equation must be formulated for every node in a system. Moreover, initial conditions must be formulated for every river section

$$p_i(x,0) = P_i(x), \quad x \in [0,\ L_i] \tag{75}$$

Differential equations describing reservoirs are straightforward:

$$\frac{dx_i}{dt} = r_i - u_i \tag{76}$$

where:

x_i – denotes amount of water stored in i-th reservoir,

r_i – denotes predicted inflow to the reservoir,

u_i – denotes the outflow from the reservoir.

In order to apply the DIDAS methodology it is necessary to formulate the equivalent linear programming problem. The most efficient way of doing this is to utilize the *linearity* of system equations and the *superposition principle*.

Let $Q_w(t) \equiv Q_I(\hat{x}, t)$ denote the water flow at a selected point of the system. Because system equations are linear, the following decomposition is possible:

$$Q_w(t) = Q_w^0(t) + Q_w^a(t) + Q_w^b(t) + Q_w^c(t) \tag{77}$$

where:

$Q_w^0(t)$ – is the solution of the system equations with $u_a(t) \equiv 0$, $u_b(t) \equiv 0$, $u_c(t) \equiv 0$, nonzero initial conditions and $q_0(t) \not\equiv 0$,

$Q_w^a(t)$ is the solution of the system equations with zero initial conditions, $u_a(t) \not\equiv 0$, $u_b(t) \equiv 0$, $u_c(t) \equiv 0$ and $q_0(t) \equiv 0$,

$Q_w^b(t)$ is the solution of the system equations with zero initial conditions, $u_a(t) \equiv 0$, $u_b(t) \not\equiv 0$, $u_c(t) \equiv 0$ and $q_0(t) \equiv 0$,

$Q_w^c(t)$ is the solution of the system equations with zero initial conditions, $u_a(t) \equiv 0$, $u_b(t) \equiv 0$, $u_c(t) \not\equiv 0$ and $q_0(t) \equiv 0$.

In the above formulas

$$Q_w^\eta(t) \equiv Q_I^\eta(\hat{x}, t) \quad \eta = \{a, b, c\} \tag{78}$$

Let us assume that the control functions $u_a(t)$, $u_b(t)$ and $U_c(t)$ can be considered as a piecewise constant functions. This is a reasonable assumptions since dispatcher of the reservoir performs his interventions only in discrete points of time (usually not more frequently than every one hour) keeping the outflow from the reservoir constant between these interventions. Therefore, the following formula can be applied

$$u_\eta(t) = \sum_{k=1}^{N-1} u_\eta(t_k)\Phi_k(t) \quad \eta = \{a, b, c\} \tag{79}$$

where

$$\Phi_k(t) = \begin{cases} 1 & \text{if } t \in [t_k, t_{k+1}] \\ 0 & \text{if } t \notin [t_k, t_{k+1}] \end{cases} \tag{80}$$

Under the above assumption, function Q_w^η can be expressed in the following way

$$Q_w^\eta(t) = \sum_{k=1}^{N-1} u_\eta(t_k)Q_{wk}^\eta(t), \quad \eta = \{a, b, c\} \tag{81}$$

where $Q_{wk}^{\eta}(t)$ is the solution of system equations under the same conditions as used to calculate Q_w^{η}, but with

$$u_{\eta}(t) \equiv \Phi_k(t) \tag{82}$$

From equation (80) follows, that each control action can be represented as a vector

$$\mathcal{U}_{\eta} \equiv [u_{\eta}(t_0), \ldots, u_{\eta}(t_{N-1})] \equiv [u_{\eta 0}, \ldots, u_{\eta(N-1)}] \tag{83}$$

Usually, we are interested only in values of Q_w^{η} in discrete moments of time, in most cases the same in which the control action is changed. Therefore, it is sufficient to compute a vector

$$\mathcal{Q}_{\eta} \equiv [Q_w^{\eta}(t_0), \ldots, Q_w^{\eta}(t_{N-1})] \equiv [Q_{w0}^{\eta}, \ldots, Q_{w(N-1)}^{\eta}] \tag{84}$$

In this situation, relation between vectors $\mathcal{U}_{\eta}$ and $\mathcal{Q}_{\eta}$ can be represented as a matrix

$$\mathcal{Q}_{\eta} = \mathcal{A}_{\eta} \mathcal{U}_{\eta} \tag{85}$$

Elements of the matrix $\mathcal{A}_{\eta}$ can be easily computed solving numerically equations describing the system. One of the known grid methods can be used for this purpose (Press at al., 1989). Knowing the matrices $\mathcal{A}_{\eta}$ the following formula can be used to calculate the response of the system for given control actions

$$\mathcal{Q}_w = \mathcal{Q}_w^0 + \mathcal{A}_a \mathcal{U}_a + \mathcal{A}_b \mathcal{U}_b + \mathcal{A}_c \mathcal{U}_c \tag{86}$$

The same principle can be applied for discretizing the reservoir equation what leads to the following matrix expression

$$\mathcal{X}_{\eta} = \mathcal{X}_{\eta}^0 + \mathcal{B}_{\eta} \mathcal{U}_{\eta} - \mathcal{C}_{\eta} \mathcal{R}_{\eta} \tag{87}$$

It should be pointed out, that the presented method of building a linear, finite dimensional equivalent of system equations is not restricted to the system with structure presented on Figure 8 but has general character and can be applied to a system described by graph of any structure, with reservoirs located anywhere between graph nodes. It is also necessary to mention, that the finite dimensional representation of a system presented in this section is *not an approximation*. The formula (86) gives the *accurate* solution of system equations, assuming that control actions belong to the class of piecewise constant functions.

Decision problem

As it has been mentioned previously, the goal of system operator (or operators) is to release water from reservoir to minimize height of peaks of flood waves in control sections of the river network.

There exist set of standard rules for operating a reservoir. Applying these rules, simulation of system behavior can be performed. This action will result in a *standard trajectory*. As a result of this simulation two situations can occur:

- The standard trajectory is *admissible*, i.e. it does not violate any constraints like maximal and minimal water levels in reservoirs and maximal water flows in all points of the river network,

- The standard trajectory is *not admissible*, i.e. one of the mentioned above constraints are violated.

Both situations can be treated according to DIDAS methodology. In the first case the following question can be posed:

- Assuming that the standard trajectory is treated as *reference trajectory* what is the corresponding *Pareto trajectory* ensuring maximal improvement of the solution, i.e. minimization of flow through river network?

In the second case, the following question can be posed:

- Assuming that the standard trajectory is treated as *reference trajectory* what is the corresponding *Pareto trajectory* which provides the best approximation of the standard one?

It should be noted, that the notions of *utopia* and *nadir* solution are rather not applicable in this case. However several experiment which would lead to generalization of this notion can be suggested, their usefulness is rather limited. This is due to the fact that according to current regulations big deviations from standard policy should be avoided, and if necessary, these deviations should be well justified. Therefore, it is reasonable to consider the standard trajectory as an equivalent of *ideal solution*.

Evidently, if solutions corresponding to the mentioned above procedure are not satisfactory for decision maker, he can perform further modifications of aspiration trajectory and repeat computations. Usually several other criteria must be taken into account by decision maker. These include required water level in reservoirs, shape of the control trajectory, etc. All these criteria can be easily included into the scheme presented above.

Implementation

The decision problem presented in the previous section must be transformed to a form which will allow applying the DIDAS formalism. In the simplest formulation of the decision problem the only performance criteria will be values of water flow in selected (critical) sections of the river network observed in discrete points of time. Therefore, the following Multiple Criteria Decision Problem (MCDM) can be formulated:

$$\min \mathcal{Q}_w \qquad (88)$$

$$\mathcal{Q}_w = \mathcal{Q}_w^0 + \mathcal{A}_a \mathcal{U}_a + \mathcal{A}_b \mathcal{U}_b + \mathcal{A}_c \mathcal{U}_c \quad w \in \mathcal{W} \qquad (89)$$

where $\mathcal{W}$ denotes set of all critical section of the river network, according to the following constraints

$$\mathcal{X}_\eta = \mathcal{X}_\eta^0 + \mathcal{B}_\eta \mathcal{U}_\eta - \mathcal{C}_\eta \mathcal{R}_\eta \qquad (90)$$

$$\mathcal{U}_{min}^{\eta} \leq \mathcal{U}_{\eta} \leq \mathcal{U}_{max}^{\eta} \tag{91}$$

$$\mathcal{X}_{min}^{\eta} \leq \mathcal{X}_{\eta} \leq \mathcal{X}_{max}^{\eta} \tag{92}$$

$$\eta = \{a, b, c\} \tag{93}$$

In the above linear programming problem $\mathcal{U}_{\eta}$ denote decision variables. This problem can be easily converted to standard DIDAS form by applying the achievement function discussed previously.

In the alternative formulation, when the decision maker is interested only in peak flow but not in shape of trajectory, the following MCDM problem can be formulated:

$$\min \xi_w \tag{94}$$

where

$$\xi_w \succeq Q_{wi} \quad i \in [0 \ldots N-1] \quad w \in \mathcal{W} \tag{95}$$

according to the constraints (89), (90), (91) and (92)

Computer implementation of the system presented in this section consists of several functional blocks which provide the following functions:

- *Model manipulation.* This part of the system provides tools for definition of structure of the river network, marking critical sections and decison variables, identification of system parameters (i.e. functions $C_i(x)$ and $D_i(x)$) for each segment of river network as well as computation of all necessary elements of finite dimensional approximation (matrices $\mathcal{A}_{\eta}$). These matrices, together with a graph describing structure of the system are converted to the corresponding linear programming problem. This part of the system includes also hydrological data base which is necessary for model identification and for performing prediction of the water inflow to the river network.

- *Interaction with the system.* This part of the system allows specification of scenarios (i.e. expected water inflow to the river network and initial conditions) as well as specification of aspiration and reservation trajectories. Moreover, the system provides tools for efficient graphical presentation of trajectories and other data.

- *Optimization and simulation.* This function is necessary for performing ordinary *what–if* analysis as well as for computing the system response corresponding to specified aspiration and reservation levels.

It is necessary to point out that computation of matrices $\mathcal{A}_{\eta}$, what requires time consuming simulation of system equations must be performed only once, after identification of system parameters. Therefore, this task can be performed off-line and its complexity will not influence the response time during decision making phase.

The discretization procedure is also memory preserving. This is due to the fact that it is not necessary to store the linear programming matrix and all elements of matrices

$\mathcal{A}_\eta$. Every element of the linear programming matrix can be computed utilizing the fact that the function $Q^\eta_{wk}(t)$ in the formula (81) has the following property

$$Q^\eta_{wk}(t) = \begin{cases} Q^\eta_{w(k-1)}(t - \tau) & \text{if } t \geq k\tau \\ 0 & \text{if } t < k\tau \end{cases} \tag{96}$$

where $\tau = t_{k+1} - t_k$. Due to this property it is necessary to compute and store in memory instead of all functions $Q^\eta_{wk}(t)$ only one function corresponding to $k = 1$.

Sufficiently efficient algorithm for solving the corresponding linear programming problem must be applied. Dimensionality of the corresponding linear programming problem can be large, however much smaller than in a case of more conventional schemes of discretization of system's equations. In this case the only decision variables of linear programming problem are components of control vectors $\mathcal{U}_\eta$ and auxiliary variables necessary to convert the multiple criteria problem to a standard single criterion one.

7 References

Anthonisse, J.M., J.K. Lenstra and M.W.P. Savelsbergh (1988). Behind the Screen: Decision Support Systemfrom and OR Point of View. Decision Support Systems, Vol. 4, pp. 413-419.

Ariav, G. and M.J. Ginzberg (1985). DSS Design: A Systemic View of Decision Support. Communications of the ACM, Vol. 28, No. 10, October 1985.

Arrow, K.J. and H. Raynaud (1986). Social Choice and Multicriterion Decision–Making. The MIT Press, 1986.

Ayati, M.B. (1987). A Unified Perspective on Decision Making and Decision Support Systems. Information Processing & Management, Vol. 23, No. 6, pp. 616–628.

Belton, V. and S.P. Vickers (1988). V.I.S.A. – VIM for MCDM. Paper presented at the VIIIth International Conference on MCDM, Manchester, England, August 1988.

Blanning, R.W. (1983). Whai is Happening in DSS? INTERFACES, Vol. 13, October 1983, pp. 71-80.

Campbell, R. and L. Sowden (1985). Paradoxes of Rationality and Cooperation: Prisoner's Dilemma and Newcomb's Problem. The University of British Columbia Press, Vancouver 1985.

Carlson, S. and C.B. Stabel (1986). Spreadsheet Programs and Decision Support: A Keystroke-Level Model of System Use. In: E. McLean and H.G. Sol. Eds: Decision Support Systems: A Decade in Perspective, Proceedings of the IFIP WG 8.3 Working Conference on Decision Support Systems, Noordwijkerhout, The Netherlands.

Charnes, A. and W. Cooper, (1975). Goal programming and multiple objective optimization. J. Oper. Res. Soc., Vol. 1, pp. 39-54.

DeSanctis, G. (1984). Computer Graphics as Decision Aids: Directions for Research. *Decision Science*, Vol. 15, pp. 463 – 487.

DeSanctis, G. and R.B. Gallupe (1987). A Foundation for the Study of Group Decision Support Systems. Management Science, Vol. 33, No. 5, May 1987.

Dickson, G.W., G. DeSanctis and D.J. McBride (1986). Understanding the Effectiveness of Computer Graphics for Decision Support: A Cumulative Experimental Approach. *Communications of the ACM*, Vol. 29, No. 1, January 1986.

Dinkelbach, W. (1982). Entscheidungsmodelle, Walter de Gruyter, Berlin, New York.

Dobrowolski, G. and M. Zebrowski (1987). Ranking and Selection of Chemical Technologies: Application of SCDAS Concept. In: A. Lewandowski and A. Wierzbicki, Eds., Theory, Software and Testing Examples for Decision Support Systems, WP–87–26, International Institute for Applied Systems Analysis, Laxenburg, Austria.

Dolk, D.R. (1986). A Generalized Model Management System for Mathematical Programming. ACM Transacions on Mathematical Software, Vol. 12, No. 2, June 1986.

Erlandson, F. E. (1981). The satisficing process: A new look. IEEE Trans. on Systems, Man and Cybernetics, Vol. SMC-11, No. 11, November 1981.

Ester, J. (1987). Systemanalyse und mehrkriterielle Entscheidung. VEB Verlag Technik, Berlin, 1987.

Fandel, G. (1972). Optimale Entscheidung bei mehrfacher Zielsetzung. Springer, Berlin Heidelberg New York, Lecture Notes in Economic and Mathematical Systems, Vol. 76.

Fishburn, P.C. (1964). Decision and Value Theory. Wiley, New York, 1964.

Fisher, W. F. (1979). Utility models for multiple objective decisions: Do they accurately represent human preferences? Decision Sciences, Vol. 10, pp. 451-477.

Gabriel, K.R. (1971). The Biplot Graphic Display of Matrices with Application to Principal Component Analysis. *Biometrika*, Vol. 58, No. 3, pp. 453–467.

Goicoechea, A., D.R. Hansen adn L. Duckstein (1982). Multiobjective Decision Analysis With Engineering and Business Applications. J. Wiley & So., 1982.

Gorecki, H., J. Kopytowski, T. Rys and M. Zebrowski (1983). A multiobjective procedure for project formulation–design of a chemical installation. In M. Grauer and
A.P. Wierzbicki, editors: Interactive Decision Analysis, Springer Verlag, Berlin,
1983.

Grauer, M., A. Lewandowski and L. Schrattenholzer (1982). Use of the reference level
approach for the generation of efficient energy supply strategies. WP-82-19, International Institute for Applied Systems Analysis, Laxenburg, Austria, 1982.

Grauer, M., and J.-L. Brillet (1982). About the portability of the DIDAS Package
(an IBM Implementation). CP-82-04, International Institute for Applied Systems
Analysis, Laxenburg, Austria, 1982.

Grauer, M. and E. Zalai (1982). A Reference Point Approach to Nonlinear Macroeconomic Planning, WP-82-134, International Institute for Applied Systems Analysis,
Laxenburg, Austria, 1982.

Grauer, M., A. Lewandowski and A.P. Wierzbicki (1983). DIDAS–theory, implementation and experience. In M. Grauer and A.P. Wierzbicki, editors: Interactive Decision
Analysis, Springer Verlag, Berlin, 1983.

Grauer, M. and S. Kaden (1984). A Nonlinear Dynamic Interactive Decision Analysis
and Support System (DIDAS/N) Users Guide, WP-84-23, International Institute
for Applied Systems Analysis, Laxenburg, Austria, 1984.

Gray, P. (1986). Group Decision Support Systems. In: E. McLean and H.G. Sol.
Eds: Decision Support Systems: A Decade in Perspective, Proceedings of the IFIP
WG 8.3 Working Conference on Decision Support Systems, Noordwijkerhout, The
Netherlands.

Hartwih, F. and B.E. Dearing (1989). Exploratory Data Analysis. Sage University Paper
No. 16, SAGE Publications.

van Hee, K.M. and A. Lapinski (1988). OR and AI Approaches to Decision Support
Systems. Decision Support Systems, Vol.4 pp. 447-459.

Hiltz, S.R and M. Turoff (1985). Structuring Computer–Mediated Communication Systems to Avoid Information Overload. Communications of the ACM, Vol. 28, No. 7,
July 1985.

Huber, G.P. (1984). Issues in the Design of Group Decision Support Systems. MIS
Quarterly, September 1984, pp.195–205.

Huerlimann, T. (1988). Reference Manual for the LPL Modelling Language. Report
No. 128 of the Institute for Automation and Operations Research, University of
Fribourg, Switzerland.

Ignizio, J.P. (1978). Goal programming–a tool for multiobjective analysis. Journal for Operational Research, 29, pp. 1109–1119, 1978.

Ignizio, J.P. (1982). Linear Programming in Single & Multiple–Objective Systems. Prentice Hall, 1982.

Iz, P. and T. Jelassi (1988). An Empirical Investigation of Multiobjective Techniques for Group Decision Making. Invited paper to the EURO IX/TIMS XXVIII Joint International Conference, Paris, France, July 6 – 8, 1988.

Jarke, M. (1986a). Group Decision Support Through Office Systems: Developments in Distributed DSS Technology. In: E. McLean and H.G. Sol, Eds: Decision Support Systems: A Decade in Perspective, Proceedings of the IFIP WG 8.3 Working Conference on Decision Support Systems, Noordwijkerhout, The Netherlands

Jarke, M., M.T. Jelassi and M.F. Shakun (1987). MEDIATOR: Toward a Negotiation Support System. European Journal of Operational Research, No. 3, September 1987.

Jarvenpaa, S.L. and G.W. Dickson (1988). Graphics and Managerial Decision Making: Research Based Guidelines. *Communications of the ACM*, Vol. 31, No. 6, June 1988.

Jelassi M.T. and R.A. Bauclair (1987). An Integrated Framework for Group Decision Support System Design. IRMIS Working Paper W703, Institute for Research on the Management of Information Systems, School of Business, Indiana University.

Jones, J.M. (1986). Decision analysis using spreadsheets. European Journal of Operational Research, Vol. 26, pp. 385-400.

Kaden, S. and T. Kreglewski (1986). Decision support system MINE–problem solver for nonlinear multi-criteria analysis. CP-86-5, International Institute for Applied Systems Analysis, Laxenburg, Austria, 1986.

Kalaba, R., Rasakhoo, N. and Tishler, R. (1983). Nonlinear Least Squares via Automatic Derivative Evaluation. Applied Mathematics and Computation, 12, pp. 119–137.

Kallio, M., A. Lewandowski and W. Orchard-Hays (1980). An implementation of the reference point approach for multiobjective optimization. WP-80-35, International Institute for Applied Systems Analysis, Laxenburg, Austria, 1980.

Kananen, I., P. Korhonen, J. Wallenius and H. Wallenius (1990). Multiple Objective Analysis of Input–Output Models for Emergency Management. *Operations Research*, Vol. 38, No. 2, March–April 1990.

Keen, P. G. W and Scott Morton, M. S. (1978). Decision Support Systems — An Organizational Perspective. Addison-Wesley Series on Decision Support.

Keen, P.G.W. (1986). Decision Support Systems: The Next Decade. In: E. McLean and H.G. Sol. Eds: Decision Support Systems: A Decade in Perspective, Proceedings of the IFIP WG 8.3 Working Conference on Decision Support Systems, Noordwijkerhout, The Netherlands.

Keeney, R.L. and H. Raiffa (1976). Decisions with Multiple Objectives: Preferences and Value Trade-offs. Wiley, New York, 1976.

Kopytowski, J. and M. Zebrowski (1988). MIDA – Experience in Theory, Software and Application of DSS in Chemical Industry. In: A. Lewandowski and A. Wierzbicki, Eds., Theory, Software and Testing Examples for Decision Support Systems, WP-88-71, International Institute for Applied Systems Analysis, Laxenburg, Austria.

Korhonen, P. and J. Laakso (1986). A Visual Interactive Method for Solving the Multiple Criteria Problem. European Journal of Operational Research, Vol. 24, pp. 277–287.

Korhonen, P.J. (1987). On Using Computer Graphics for Solving MCDM Problems. In: *Towards Interactive and Intelligent Decision Support* Systems, Y. Sawaragi, K. Inoue and H. Nakayama, Eds. Proceedings of the Seventh International Conference on Multiple Criteria Decision Making, Kyoto, Japan, August 1986. Lecture Notes in Economics and Mathematical Systems, Vol. 286, Springer Verlag.

Korhonen, P. (1988). A Visual Reference Direction Approach to Solving Discrete Multiple Criteria Problems. European Journal of Operational Research, Vol. 34, pp. 152–159.

Korhonen, P. (1988). Using Harmonious Houses for Visual Pairwise Comparison of Multiple Criteria Alternatives. Working Paper F-203, Helsinki School of Economics, Helsinki, Finland, December 1988.

Kraemer, K.L. and J.L. King (1988). Computer-Based Systems for Cooperative Work and Group Decision Making. ACM Computing Surveys, Vol. 20, No. 2, June 1988.

Kreglewski, T. and Lewandowski, A. (1983). MM–MINOS – an Integrated Interactive Decision Support System. CP-83-63, International Institute for Applied Systems Analysis, Laxenburg, Austria.

Kreglewski, T., T. Rogowski, A. Ruszczynski, J. Szymanowski (1984). Optimization methods in FORTRAN, PWN, Warsaw, 1984 (in Polish).

Kreglewski T., J. Paczynski, J. Granat, A. P. Wierzbicki (1988). IAC–DIDAS–N A Dynamic Interactive Decision Analysis and Support System for Multicriteria Analysis of Nonlinear Models with Nonlinear Model Generator supporting model analysis. WP-88-112, International Institute for Applied Systems Analysis, Laxenburg, Austria.

Kreglewski, T. (1988). Nonlinear Optimization Techniques in Decision Support Systems. In: A. Lewandowski, A.P. Wierzbicki, Eds., *Theory, Software and Testing Examples in Decision Support Systems*. WP–88–071, International Institute for Applied Systems Analysis, Laxenburg, Austria.

Kruskal, J.B. and M. Wish (1989). Multidimensional Scaling. Sage University Paper No. 11, SAGE Publications.

Lewandowski, A. and M. Grauer (1982). The reference point optimization – methods of efficient implementation. In: *Multiobjective and Stochastic Optimization*, M. Grauer, A. Lewandowski and A.P. Wierzbicki, Eds., CP–82–S12, International Institute for Applied Systems Analysis, Laxenburg, Austria.

Lewandowski, A. (1982). A Program Package for Linear Multiple Criteria Reference Point Optimization–Short User Manual, WP-82-80, International Institute for Applied Systems Analysis, Laxenburg, Austria, 1982.

Lewandowski, A., T. Rogowski and T. Kreglewski (1984). Application of DIDAS methodology to flood control problems – numerical experiments. In M. Grauer, M. Thompson, A.P. Wierzbicki, editors: Plural Rationality and Interactive Decision Processes, Proceedings, Sopron 1984, Springer Verlag, Berlin.

Lewandowski, A. and T. Kreglewski (1985). A nonlinear version of DIDAS system, Collaborative volume: Theory, Software and Test Examples for Decision Support Systems, International Institute for Applied Systems Analysis, Laxenburg, Austria, 1985.

Lewandowski, A., S. Johnson and A.P. Wierzbicki (1986). A Selection Committee Decision Support System: Implementation, Tutorial Example and Users Manual. International Institute for Applied Systems Analysis, Laxenburg, Austria, 1986; presented also at the MCDM Conference in Kyoto, Japan, August 1986.

Lewandowski, A. (1986). Problem Interface for Nonlinear DIDAS. WP-86-50, International Institute for Applied Systems Analysis, Laxenburg, Austria.

Lewandowski, A. (1987). Selection Committee Decision Analysis and Support Systems (SCDAS) – Users Manual V. 2.0. International Institute for Applied Systems Analysis, Laxenburg, Austria, manuscript.

Lewandowski, A. and A.P. Wierzbicki (1987). Interactive Decision Support Systems – The Case of Discrete Alternatives for Committee Decision Making. WP–87–38, International Institute for Applied Systems Analysis, Laxenburg, Austria.

Lewandowski, A. (1988a). SCDAS – Decision Support System for Group Decision Making: Short User's Manual. International Institute for Applied Systems Analysis, Laxenburg, Austria, manuscript.

Lewandowski, A. (1988b). Distributed SCDAS – Decision Support System for Group Decision Making: Functional Specification. International Institute for Applied Systems Analysis, Laxenburg, Austria, manuscript.

Lewandowski, A. and A.P. Wierzbicki (1988c). Theory, Software and Testing Examples in Decision Support Systems. WP–88–71, International Institute for Applied Systems Analysis, Laxenburg, Austria.

Lewandowski, A. and A.P. Wierzbicki (1988d). Decision Support Systems Using Reference Point Optimization. In: A. Lewandowski, A.P. Wierzbicki, Eds., *Theory, Software and Testing Examples in Decision Support Systems*. WP–88–071, International Institute for Applied Systems Analysis, Laxenburg, Austria.

Lewandowski, A., T. Kreglewski, T. Rogowski and A.P. Wierzbicki (1988e). Decision Support Systems of DIDAS Family (Dynamic Interactive Decision Analysis and Support). In: A. Lewandowski, A.P. Wierzbicki, Eds., *Theory, Software and Testing Examples in Decision Support Systems*. WP–88–071, International Institute for Applied Systems Analysis, Laxenburg, Austria.

Lewandowski, A. (1988). Short Software Descriptions. WP–88–109, International Institute for Applied Systems Analysis, Laxenburg, Austria.

Lucas, H.C. (1981). An Experimental Investigation of the Use of Computer–Based Graphics in Decision Making. *Management Science*, Vol. 27, No. 7, July 1981.

Luce, R.D. and H. Raiffa, (1957). Games and decisions. Wiley, New York.

Majchrazk, J. (1981). The implementation of the multicriteria reference point optimization approach to the Hungarian Regional Investment Allocation Model. WP–81–154, International Institute for Applied Systems Analysis, Laxenburg, Austria.

Makowski, M. and J. Sosnowski (1984). A decision support system for planning and controlling agricultural production with a decentralized management structure. In M. Grauer, M. Thompson, A.P. Wierzbicki, editors: Plural Rationality and Interactive Decision Processes, Proceedings, Sopron 1984, Springer Verlag, Berlin.

Makowski, M. and J. Sosnowski (1988). User Guide to a Mathematical Programming Package for Multicriteria Dynamic Linear Problems HYBRID Version 3.1, WP-88-111, International Institute for Applied Systems Analysis, Laxenburg, Austria, December 1988.

Makowski, M. and J.S. Sosnowski Mathematical Programming Package for Multicriteria Dynamic Linear Problems HYBRID. Methodological Guide to Version 3.1. In: A. Lewandowski, A.P. Wierzbicki, Eds., *Theory, Software and Testing Examples in Decision Support Systems*. WP–88–071, International Institute for Applied Systems Analysis, Laxenburg, Austria.

Marechal, B. and J.-P. Brans (1988). Geometrical Representation for MCDA. *European Journal of Operational Research*, Vol. 34, pp. 66–77.

Masud, H.S. and C.L. Hwang, (1981). Interactive sequential goal programming. Jour. Oper. Res. Soc., Vol. 32, pp. 391-400.

Messner, S. (1985). Natural gas trade in Europe and interactive decision analysis, In G. Fandel, M. Grauer, A. Kurzanski and A.P. Wierzbicki, eds., Large-Scale Modelling and Interactive Decision Analysis, Proceedings Eisenach, Springer Verlag, Berlin, 1985.

Messner, S. (1985). User's guide for the matrix generator of MESSAGE-II. Part 1: model description. WP-84-071, International Institute for Applied Systems Analysis, Laxenburg, Austria.

Mirkin B.G. (1979). Group Choice. John Willey & Sons, New York.

Murtagh, B.A. and M.A. Saunders (1980). MINOS/Augmented, Technical Report, SO1-80-14, Systems Optimization Laboratory, Stanford University, 1980.

Murtagh, B.A. (1981). Advanced Linear Programming: Computation and Practice. McGraw–Hill, Inc., 1981.

Nakayama, H. and Y. Sawaragi (1983). Satisficing trade-off method for multiobjective programming. In M. Grauer and A.P. Wierzbicki, editors: Interactive Decision Analysis, Springer Verlag, Berlin, 1983.

Naylor, T.H. (1982). Decision Support Systems or Whatever Happened to M.I.S.? INTERFACES, Vol. 12, No.4, August 1982, pp. 92-94.

Ogryczak, W., K. Studzinski, K. Zorychta (1988). Dynamic Interactive Network Analysis System – DINAS, version 2.1. User's Manual. WP-88-114, International Institute for Applied Systems Analysis, Laxenburg, Austria.

Ogryczak, W., K. Studzinski and K. Zorychta (1988). A Generalized Reference Point Approach to Multiobjective Transshipment Problem with Facility Location. In: A. Lewandowski, A.P. Wierzbicki, Eds., *Theory, Software and Testing Examples in Decision Support Systems*. WP-88-071, International Institute for Applied Systems Analysis, Laxenburg, Austria.

Ogryczak, W., K. Studzinski and K. Zorychta (1988). Solving Multiobjective Distribution – Location Problems with the DINAS System. In: A. Lewandowski, A.P. Wierzbicki, Eds., *Theory, Software and Testing Examples in Decision Support Systems*. WP-88-071, International Institute for Applied Systems Analysis, Laxenburg, Austria.

Paczynski, J. (1988). Nonlinear Computer Models — Issues of Generation and Differentiation. In: A. Lewandowski, A.P. Wierzbicki, Eds., *Theory, Software and Testing Examples in Decision Support Systems*. WP–88–071, International Institute for Applied Systems Analysis, Laxenburg, Austria.

Parker, B. J. and Al-Utabi, G. A. (1986). Decision support systems: The reality that seems to be hard to accept? OMEGA Int. Journal of Management Science, Vol. 14, No. 2, 1986.

Pearson, M.L. and J.E. Kulp (1981). Creating an Adaptive Computerized Conferencing System on UNIX. In: R.P. Uhlig, Ed., Computer Message Systems, North–Holland, 1981.

Rappoport, A. (1984). The uses of experimental games. In: Grauer, M., Thompson, M., Wierzbicki, A. P. Eds: Plural Rationality and Interactive Decision Processes. Proceedings, Sopron, Hungary, 1984. Lecture Notes in Economics and Mathematical Systems, Vol. 248. Springer-Verlag, Berlin.

Rogowski, T., J. Sobczyk, A. P. Wierzbicki (1988). IAC–DIDAS–L Dynamic Interactive Decision Analysis and Support System, Linear version. WP-88-110, International Institute for Applied Systems Analysis, Laxenburg, Austria.

Rogowski, T. (1988). Dynamic Aspects of Multiobjective Trajectory Optimization in Decision Support Systems. In: A. Lewandowski, A.P. Wierzbicki, Eds., *Theory, Software and Testing Examples in Decision Support Systems*. WP–88–071, International Institute for Applied Systems Analysis, Laxenburg, Austria.

Sakawa, M. (1983). Interactive fuzzy decision making for multiobjective nonlinear programming problems. In M. Grauer and A.P. Wierzbicki, editors: Interactive Decision Analysis, Springer Verlag, Berlin, 1983.

Sawaragi, Y., H. Nakayama and T. Tanino (1985). Theory of Multiobjective Optimization. Academic Press.

Schwartz, T. (1986). The Logic of Collective Choice. Columbia University Press, New York, 1986.

Simkin, D. and R. Hastie (1987). An Information–Processing Analysis of Graph Perception. *Journal of the American Statistical Association*, Vol. 82, No. 398, June 1987.

Simon, H.A. (1957). Models of Man. Macmillan, New York, 1957.

Sprague, R. H. and Carlson, C. Eds. (1982). Building Effective Decision Support Systems. Prentice Hall, Inc.

Spronk, J. (1981). Interactive Multiple Goal Programming: Applications to Financial
    Planning. Martinus Nijhoff Publishing, 1981.

Stadtler, H., M. Groeneveld and H. Hermannsen (1988). A comparison of LP software
    on personal computers for industrial applications. European Journal of Operational
    Research, Vol. 35, pp. 146-159.

Stefik, M., D.G. Bobrow, G. Foster, S. Lanning, and D. Tatar (1987). WYSIWIS Re-
    vised: Early Experiences with Multiuser Interfaces. ACM Transactions on Office
    Information Systems, Vol. 5, No. 2, pp. 147-167.

Steuer, R. and E.V. Choo (1983). An interactive weighted Chebyshev procedure for
    multiple objective programming. Mathematical Programming, 26, pp. 326-344,
    1983.

Strubegger, M. (1985). An approach for integrated energy-economy decision analysis:
    the case of Austria. In G. Fandel, M. Grauer, A. Kurzanski and A.P. Wierzbicki,
    eds., Large-Scale Modelling and Interactive Decision Analysis, Proceedings Eise-
    nach, Springer Verlag, Berlin, 1985.

Thierauf, R.J. (1982). Decision Support Systems for Effective Planning and Control.
    Prentice Hall, 1982.

Thierauf, R.J. (1988). User–Oriented Decision Support Systems: Accent On Problem
    Solving. Prentice Hall, 1988.

Tufte, E.R. (1983). The Visual Display of Quantitative Information. Graphic Press,
    Chestire, Connecticut.

Wierzbicki, A.P. (1975). Penalty methods in solving optimization problems with vector
    performance criteria. VI Congress of IFAC, Boston 1975.

Wagner, G.R. (1981). Decision Support Systems: The Real Substance. INTERFACES,
    Vol. 11, No. 2, April 1981, pp. 77-86.

Watson, H.J. and M.M. Hill (1983). Decision Support Systems or What Didn't Happen
    with MIS. INTERFACES, VOl. 13, October 1983, pp. 91-88.

Weistroffer, H.R. (1984). A combined over- and under-achievement programming ap-
    proach to multiple objectives decision making. Large Scale Systems, Vol. 7, pp. 47-
    58.

Wierzbicki, A.P. (1977). Basic properties of scalarizing functionals for multiobjective
    optimization. Mathematische Operationsforschung und Statistik, Ser. Optimization
    8, Nr 1, 1977.

Wierzbicki, A.P. (1980). The use of reference objectives in multiobjective optimization. In G. Fandel and T. Gal, eds., Multiple Criteria Decision Making, Theory and Applications, Springer Verlag, Heidelberg 1980.

Wierzbicki, A.P. (1982). A mathematical basis for satisficing decision making. Mathematical Modelling, 3, pp. 391–405, 1982.

Wierzbicki, A. P. (1983a). Negotiation and mediation in conflicts: The role of mathematical approaches and methods. In: Chestnut, H. et al., Eds: Supplemental Ways to Increase International Stability. Pergamon Press, Oxford, 1983.

Wierzbicki, A.P. (1984a). Negotiation and mediation in conflicts, II: Plural rationality and interactive decision processes. In M. Grauer, M. Thompson, A.P. Wierzbicki, editors: Plural Rationality and Interactive Decision Processes, Proceedings, Sopron 1984, Springer Verlag, Berlin.

Wierzbicki, A. P. (1985). Negotiation and mediation in conflicts: Plural rationality and interactive decision processes. In: Grauer, M., Thompson M. and Wierzbicki A. P., Eds: Plural Rationality and Interactive Decision Processes, Proceedings, Sopron, 1984. Lecture Notes in Economics and Mathematical Systems, Vol. 248. Springer Verlag, Berlin.

Wierzbicki, A.P. (1986). On the completeness and constructiveness of parametric characterizations to vector optimization problems. OR-Spektrum, 8, pp. 73–87, 1986.

Wynne, B. (1982). Decision support systems — a new plateau of opportunity or more emperor's clothing? INTERFACES, Vol. 12, No. 1, February 1982. Rietveld, P., (1980). Multiple objective decision methods and regional planning. North-Holland, Amsterdam.

Vazsonyi, A. (1982). Decision Support Systems, Computer Literacy and Electronic Models. INTERFACES, Vol. 12, No. 1, February 1982, pp. 74-78.

Venkatraman, S. S. (1989). DSS: Just an alias for MIS? COMPUTER PERSONNEL, A Quarterly Publication of the Special Interest Group on Computer Personnel Research, Vol. 12, No. 2, December 1989.

Zeleny, M. (1982). Multiple criteria decision making. McGraw-Hill, New York.

```
(**************************************************

    A simple linear programming model with three variables

    x,y and z, two restrictions and a minimizing

    objective function written in LPL.

(***************************************************)



    {$Cv(1,1)           compiler directive}

    {$Rv(1,5)+n         compiler directive}



{Variable Declaration}

var

  x y z



{Model section}

model

  Restriction1: x + 43*y = 4;

  Restriction2: y - (40+5)*z <= 5;

  Objective:    x + y - z = MAX(costs);

end
```

Figure 4: Sample model specification in LPL language

```
NAME            model1

ROWS

 E   RESTR0

 L   RESTR1

 N   OBJEC2

COLUMNS

        X         RESTR0       1

        X         OBJEC2       1

        Y         RESTR0       43

        Y         RESTR1       1

        Y         OBJEC2       1

        Z         RESTR1       -45

        Z         OBJEC2       -1

RHS

     ..RHS        RESTR0       4

     ..RHS        RESTR1       5

ENDATA
```

Figure 5: MPS file generated by LPL program

```
vardec v, d, f;

varobj SI, SR, TL, MRR;

varcon cv, cf, cd;

var Lv, Lf, Ld;

begin

   Lv:=ln(v); Lf:=ln(1000*f); Ld:=ln(1000*d);

   SI:=-4.13+0.92*Lv-0.16*Lf+0.43*Ld;   (* definition of objectives *)

   SR:= 7.49-0.44*Lv+1.16*Lf-0.061Ld;

   TL:= 21.90-1.94*Lv-0.30*Lf-1.04*Ld;

   MRR:= -11.33+Lv+Lf+Ld;

   cv:=v; cf:=f; cd:=d;                 (* definition of constraints *)

end.
```

Figure 6: Definition of the nonlinear model in extended Pascal

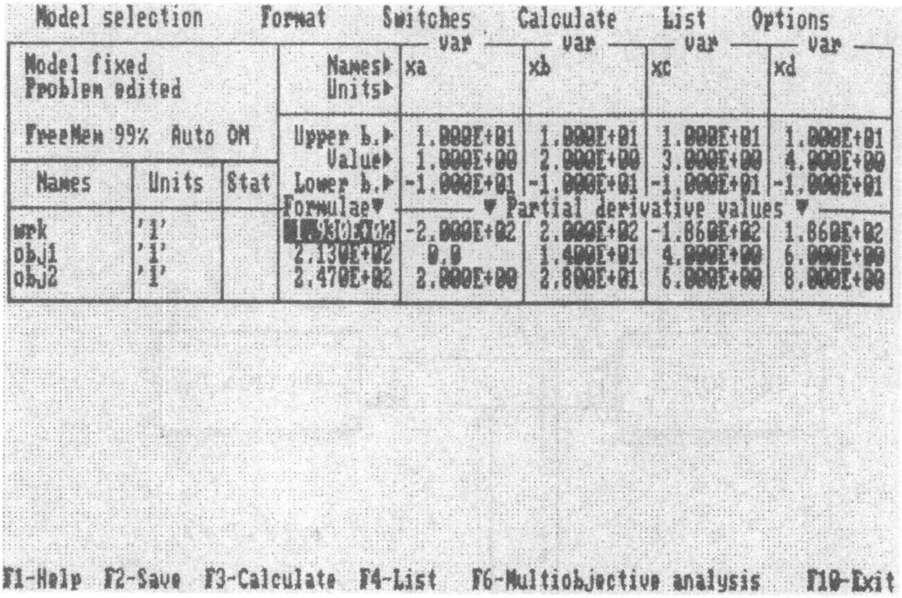

Model selection	Format	Switches	Calculate	List	Options

Model fixed Problem edited	Names▶ Units▶	var xa	var xb	var xc	var xd
FreeMem 99% Auto ON	Upper b.▶ Value▶	1.000E+01 1.000E+00	1.000E+01 2.000E+00	1.000E+01 3.000E+00	1.000E+01 4.000E+00

Names	Units	Stat	Lower b.▶ Formulae▼	-1.000E+01	-1.000E+01	-1.000E+01	-1.000E+01
				▼ Partial derivative values ▼			
wrk	'1'		█████████	-2.000E+02	2.000E+02	-1.860E+02	1.860E+02
obj1	'1'		2.130E+02	0.0	1.400E+01	4.000E+00	6.000E+00
obj2	'1'		2.470E+02	2.000E+00	2.800E+01	6.000E+00	8.000E+00

F1-Help F2-Save F3-Calculate F4-List F6-Multiobjective analysis F10-Exit

Figure 7: Definition of the nonlinear model in IAC–DIDAS–N system

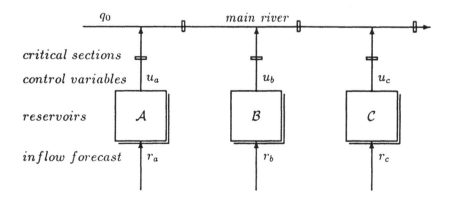

Figure 8: Structure of the river network

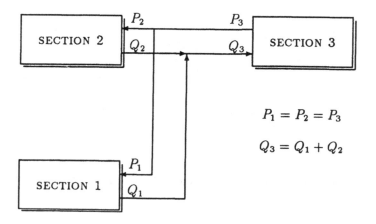

Figure 9: Structure of the node

INTERACTIVE MULTI-OBJECTIVE PROGRAMMING
AND ITS APPLICATIONS

H. Nakayama
Konan University, Kobe, Japan

ABSTRACT:

Many practical problems often have several objectives conflicting with
each other, and we need to make the balanced decision from the total view
point. For these problems, the traditional mathematical programming is
not valid, and instead the multi-objective programming have been developed. Among them, the aspiration level approach has been widely recognized to be effective in many practical fields.

In this paper, some of techniques based on aspiration levels are
discussed along with a device for the automatic trade-off using parametric optimization techniques. Some practical examples show that the
methods are user-friendly and 'synayakana' in Japanese (i.e., flexible
and robust, in English) to the multiplicity of value judgement.

1. Introduction

The aim of decision support systems (DSS) is not to automatize decision making. Without existence of decision makers, we can not discuss DSS. Therefore, we need to consider DSS not only as the information process using traditional mathematical science, but also as a total system including human beings.

Above all, the one of most important and difficult things in decision making is to make a decision under circumstances changing very often. Therefore, DSS should be 'flexible' or 'adaptable' or 'self-organizing' for such changes. The Japanese word 'synayakana' covers all 'flexible', 'adaptable' and 'self-organizing'. Among factors of circmstance in decision making, the uncertainty of environment and the multiplicity of value judgment are very important. In particular, this report focuses on the muitiplicity of value judgment, and suggest a trial of DSS behaving in a 'synayakana' way for it.

In the process of decision making, there are three modellings, namely, the structure model, the impact model and the evaluation model.

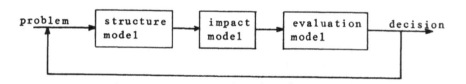

Fig. 1   Three models in decision making process

By the structure model, we mean to make a model to know the structure of the problem: what the problem is, which factors composit the problem, how they interrelate and so on. Through the process, the objective of the problem and alternatives to perform it are specified. Therefore, several kinds of factor analysis of the problem are made at this stage.

In the impact model, we analyze how much impact is given on each objective by performing an alternative from viewpoints of physical, chemical, biological, socio-economical and so on. Therefore, the cause-effect analysis is made at this stage.

In a traditional sense of modelling, the above two models are mainly the subject to be considered. However, depending on individuals and situations, different evaluations are possible even for an impact. Now,

the evaluation model appears inevitable and very important in decision
making. The decision is made on a basis of value judgment of decision
makers. In the evaluation model, therefore, it is mainly considered how
to make the information process of value judgment of decision makers. At
this stage, much attention should be paid to the human factors in
decision making.

Originally, decisions may vary depending on decision makers. Therefore, one of the most important characteristics necessary for DSS is to
help decision makers, whoever they are, and lead them to their own
decisions. In other words, DSS should be 'synayakana' for the multiplicity of value judgment. Of course, even one decision maker often changes
his value judgment during the decision process. It is needless to say
that DSS should cover such changes of value judgment of a decision maker
too.

From the above consideration, we can see that it is not so easy to
automatize decision making, because it is almost impossible to include
all value judgments in a DSS. (In cases that there occur only limited
value judgments, this would be possible.) Among many kinds of intelligent information process, the value juegment is most 'human'. Even if
the technology of artificial intelligence and knowledge engineering will
be highly developed in future, the value–judgment will probably be left
'hard to deal' to the last. Therefore, it is much better to consider DSS
as a cooperative system of man and computer, in which they bear their own
share in DSS taking their strong points, rather than the automation
system of decision making which exculdes man from the system.

2. Multi-objective Programming

Some of practical problems can be formulated as the following mathematical programming:

(P)   Minimize $f(x)=(f_1(x), f_2(x),...,f_r(x))$ over $x \in X$.

The constraint set X may be represented by

$$g_j(x) \leqq 0, \qquad j=1,..,m,$$

and/or a subset of R^n itself.

The function form of f and g and the set X are identified through
the structure modelling and the impact modelling. We focus our discus-

sion on cases that such identification can be made relatively easily, as
are shown in the following examples:

1) blending .... raw materials should be mixed appropriately from
                    several viewpoints:

   (i) feed
          raw materials: corn, wheat, fish meal, etc.
          criteria: cost, neutrition, etc.
  (ii) plastic materials (Nakayama et al. 1986)
          raw materials: alminium oxide, sillica oxide, calcium oxide,
                            etc.
          criteria: cost, melting point, acid (alkaline, aqua)-resisting
                      characteristic, coefficient of thermal expansion,
                      etc.
 (iii) cement
          raw materials: limestone, clay, iron, etc.
          criteria: cost, hydraulic (silica, iron) modulus
  (iv) steel production
          raw materials: several kinds of iron stones, etc.
          criteria: cost, physical (chemical) characteristics of steel,
                      import condition of raw iron stones, etc.
   (v) portfolio (Nakayama 1989)
          raw materials: several kinds of bond and stock, etc.
          criteria: expected return, risk, growth, etc.

2) design .... design variables should be decided appropriately from
                    several viewpoints:
   (i) camera lens
          variables: interval of lenses, curvature of lenses, etc.
          criteria: cost, weight , length, focus, caliber, aberration,
                      etc.
  (ii) erection management of cable-stayed bridge (Ishido et al. 1987)
          variables: amount of shim-adjustment, etc.
          criteria: cost, deviation of cable-tension, deviation of
                      camber, etc.

3) planning ... an appropriate policy should be decided on the basis of
                    predicted time dependent factors from several view-
                    points:
   (i) scheduling of string selection in steel manufacturing (Ueno

et al. 1990)

    policies: lot formaticn (kind and size of lot), etc.

    criteria: customers' demand (specifications, due date, etc.),

            efficency of manufacturing, etc.

(ii) long term planning of atomic power plants

    policies: how many, how large and what kind of atomic power

            plants at each period, etc.

    criteria: cost, amount of power, existing amount of pultonium,

            etc.

These problems can be formulated as mathematical programming relatively easily. However, unlike traditional mathematical programming with a single objective function, an optimal solution can not be defined in general.[1] The balancing among plural objective functions is left to the value judgment of decision makers. This total balancing among criteria is usually called trade-off. It should be noted that there are very many criteria, say, over one hundred in some problems such as erection management of cable stayed bridge, and camera lens design. Therefore, it is very important to develop effective methods for helping decision makers to trade-off easily even in probllems with very many criteria.

Interactive multi-objective programming has been developed remarkably for the last about fifteen years: the methods search a solution while eliciting information on the value judgment of decision makers. Among several methods in interactive multi-objective programming, the aspiration level approach is now recognized very effective in practice, because it does not require any consistency of decision makers' judgment, and in addition aspiration level is very easy for decision makers to answer.

3. Aspiration Level Techniques for Interactive Multiobjective Programming

For simplicity for a while, suppose that X stands for all constraints including g_j ($1 \leq j \leq m$) in the previous section. In the aspiration level approach, the aspiration level at the k-th iteration $\bar{f}^k$ is modified as follows:

$$\bar{f}^{k+1} = \text{ToP}(\bar{f}^k) \tag{3.1}$$

Here, the operator P selects the Pareto solution nearest in some sense to the given aspiration level $\bar{f}^k$. The operator T is the trade-off operator which changes the k-th aspiration level $\bar{f}^k$ if the decision maker does not compromise with the shown solution $P(\bar{f}^k)$. Of course, since $P(\bar{f}^k)$ is a Pareto solution, there exists no feasible solution which makes all criteria better than $P(\bar{f}^k)$, and thus the decision maker has to trade-off among criteria if he wants to improve some of criteria. Based on this trade-off, a new aspiration level is decided as $\mathrm{ToP}(\bar{f}^k)$. Similar process is continued until the decision maker obtain an agreeable solution. This idea is implemented in DIDASS (Kallio et al. 1980 and Grauer et.al. 1984) and the satisficing trade-off method (Nakayama 1984).

The operation which gives $P(\bar{f}^k)$ from $\bar{f}^k$ is performed by some auxiliary scalar optimizatin. The objective function in this auxiliary optmization is called an achievement function in some literature (Wierzbicki 1986). Let f_i^* be an ideal value which is usually given in such a way that $f_i^* < \mathrm{Min} \ \{f_i(x) \,|\, x \varepsilon X\}$, and let f_{*i} be a nadir value which is usually given by $f_{i*} = \mathrm{Max}_{1 \leq j \leq r} \ f_i(x_j^*)$ where $x_j^* = \arg \ \mathrm{Min}_{x \varepsilon X} \ f_j(x)$.

Then, typical examples of achievement function based on Tchebyshev norm are given in the following:

$$P_1 = \mathrm{Max}_{1 \leq i \leq r} \ w_i(f_i(x) - f_i^*) \longrightarrow \mathrm{Min}$$

where

$$w_i = \frac{1}{\bar{f}_i - f_i^*}$$

or

$$P_2 = \mathrm{Max}_{1 \leq i \leq r} \ w_i(f_i(x) - \bar{f}_i) \longrightarrow \mathrm{Min}$$

where

$$w_i = \frac{1}{f_{*i} - f_i^*}$$

P_1 is used in the satisficing tradeoff method (Nakayama 1984), while P_2 is used in DIDASS (Grauer et al. 1984). The desirable properties of achievement functions were discussed in Nakayama 1985 and Wierzbicki 1986. Both of these two functions have almost the same characteristics from this point of view. However, there is a slight difference between Pareto solutions obtained by optimizing these functions (see, for example, Nakayama 1990).

Remark 3.1 It should be noted that the solution obtained by optimizing the functions P_1 or P_2 is just a weak Pareto solution. As is well known, if we want to get a strong Pareto optimal solution rather than a merely weak Pareto solution, we can use the following augmented achievement functions:

$$q_1 = \underset{1 \leq i \leq r}{\text{Max}} \; w_i(f_i(x) - f_i^*) + \varepsilon \sum_{i=1}^{r} w_i f_i(x)$$

or

$$q_2 = \underset{1 \leq i \leq r}{\text{Max}} \; w_i(f_i(x) - \bar{f}_i) + \varepsilon \sum_{i=1}^{r} w_i f_i(x)$$

Several other achievment functions are discussed in Grauer et al. 1984 and Steuer 1986.

Since the above achievement functions are not smooth, the minimization for them is usually performed by solving the equivalently transformed problem. Instead of minimizing p_2, for example, we solve the following:

(Q) Minimize z

   subject to $w_i(f_i(x) - \bar{f}_i) \leq z$

   $x \in X$.

From a standpoint of parametric optimization, p_2 is more convenient than p_1, because it changes only the values of right hand side according to the change of aspiration levels. In the following section, we shall discuss a method for the automatic trade-off.

4. Automatic Trade-off using Parametric Analysis

4.1 Trade-off

As was stated above, since the solutions obtained by the projection $P(\bar{f})$ are Pareto optimal, there is no other feasible solution that improves all objective functions. Therefore, if decision makers want to improve some of objective functions, then they have to agree with some sacrifice of other objective functions. In cases decision makers are not satisfied with the solution for $P(\bar{f}^k)$, they are requested to answer their new aspiration level $\bar{f}^{k+1}$. Let x^k denote the Pareto solution obtained by projection $P(\bar{f}^k)$, and classify the objective functions into the following three groups:

(i) the class of criteria which are to be improved more,

(ii) the class of criteria which may be relaxed,

(iii) the class of criteria which are acceptable as they are.

The index set of each class is represented by I_I^k, I_R^k, I_A^k, respectively. Clearly, $\bar{f}_i^{k+1} < f_i(x^k)$ for all $i \in I_I^k$. Usually, for $i \varepsilon I_A^k$, we set $\bar{f}_i^{k+1}$ $= f_i(x^k)$. For $i \in I_R^k$, decision makers have to agree to increase the value of $\bar{f}_i^{k+1}$. It should be noted that an appropriate sacrifice of f_j for j I_R^k is needed for getting the improvement of f_i for $i \varepsilon I_I^k$.

It is of course possible for decision makers to answer new aspiration levels of all objective functions. In practical probelms, however, we often encounter cases with very many objective functions as was stated in the previous section. Under this circumstance, decision makers tend to get tired with answering new aspiration levels for all objective functions. Usually, the feeling that decision makers want to improve some of criteria is much stronger than the one that they compromise with some compensatory relaxation of other criteria. Therefore, it is more practical in problems with very many objective functions for decision makers to answer only their improvement rather than both improvement and relaxation.

At this stage, we can use the assignment of sacrifice for f_j ($j \in I_R$) which is automatically set in the equal proportion to $\lambda_i w_i$, namely, by

$$\Delta f_j = \frac{-1}{N \lambda_j w_j} \sum_{i \varepsilon I_I} \lambda_i w_i \Delta f_i \qquad (4.1)$$

where N is the number of elements of the set I_R, and λ is the Lagrange multiplier associated with the constraints in Problem [Q]. The reason why (4.1) is available is that $(\lambda_1 w_1, \ldots, \lambda_r w_r)$ is the normal vector of the tangent hyperplane of the Pareto surface under appropriate conditions.   By doing this, in cases where there are a large number of criteria, the burden of decision makers can be decreased so much. Of course, if decision makers do not agree with this quota Δf_j laid down automatically, they can modify them in a manual way.

4.2 Exact Trade-off in Multi-objective Linear Programming

In trade-off, the amount of relaxation given by (4.1) is not sufficient in general, in particular, in nonlinear and convex cases, because the quota Δf_j is decided in such a way that the new aspiration level is merely on the supporting hyperplane for the Pareto surface. In cases where all functions in [P] are linear and X is a polyhedral set in an n-dimensional Euclidean space, however, a more precise analysis is possible by using parametric optimization techniques. In other words, parametric optimization techniques can give us the exact amount of relaxation so that the new aspiration level may be on the Pareto surface. A method along this line will be suggested in this section. In the following, we consider the following linear multi-objective programming problem:

[MOLP]   Minimize $(c_1 x, \ldots, c_r x)$

              subject to

                    $Dx \leq d$

                    $x \geq 0.$

Recall that every strong Pareto solution to [MOLP] is proper. At the first stage, therefore, we use the following augmented objective function with $\varepsilon > 0$ in the auxiliary min-max problem:

[ALP]   Minimize $z + \varepsilon \sum w_i c_i x$

              subject to $c_i x - \bar{f}_i^k \leq (1/w_i)z,$ $i=1,\ldots,r$

                    $x \in X,$

where $X := \{x \in R^n \mid Dx \leq d, \ x \geq 0\}.$

It is well known that the solution to [ALP] is strongly Pareto optimal. Moreover, in this formulation, as is seen in Korhonen et al. (1988), it is very easy to change the i-th objective function into a constraint function by replacing $1/w_i$ with 0, and vice versa.

One way to decrease the burden of decision makers in trade-off is the following:

<u>Step 1.</u> Ask decision makers the new aspiration level $\bar{f}_i^{k+1}$ for the objective functions to be improved, f_i ($i \varepsilon I_I$).

<u>Step 2.</u> Let r be the index of the objective function to be relaxed most.

<u>Step 3.</u> Let $f^k := f(x^k)$ where x^k is the solution to [ALP]. Decide the new aspiration level $\bar{f}_j^{k+1} := f^k + \Delta f_j^k$ ($j \varepsilon I_R$) for the objective function to be relaxed as follows: Set

$$\Delta f_j^k = - \sum_{i \in I_I} \lambda_i w_i \Delta f_i / (N \lambda_j w_j), \qquad j \in I_R \backslash \{r\},$$

where

$$\Delta f_i^k = \bar{f}_i^{k+1} - f_i^k, \qquad i \in I_I.$$

Decide $\bar{f}_r^{k+1}$ by solving the following linear parametric problem:

[PLP]   Minimize   z

subject to $c_i x - f_i^k - t \Delta f_i^k \leq 0, \qquad i \in I_I$

$c_j x - f_j^k - t \Delta f_j^k \leq 0, \qquad j \in I_R \backslash \{r\}$

$c_r x - f_r^k \leq (1/w_r) z$

$x \in X$

In the parametric linear programming [PLP], at first, the new base is obtained by the sensitivity analysis from the final tableau of [ALP] in which the coefficient vector of objective function is changed from $(1, \varepsilon w_1 c_1', ..., \varepsilon w_n c_n')$ into $(1, 0, ..., 0)$, and the right hand side vector $\bar{f}^k$ into f^k, and finally the column vector associated with z in the coefficient matrix of the constraint is changed from $(1/w_1, ..., 1/w_{r-1}, 1/w_r)$ into $(0, ..., 0, 1/w_r)$. In many practical cases, the new base in the above process can be obtained in one pivotting as will be shown later (Lemma 4.4).

Secondly, the solution of [PLP] with t=1 is obtained by the right hand side sensitivity analysis. At corner points of the Pareto surface with kink, the solutions are degenerate. As in the usual parametric optimization, the dual simplex method is used for getting the new base at such degenerate solutions.

Note that the obtained new aspiration level is already Pareto optimal. Therefore, since only a few pivotting are usually needed in these techniques, we can obtain the new Pareto solution associated with a new aspiration level very quickly. Moreover, we can make a microjustification around the obtained Pareto solution along the direction $(\Delta f_1^k, ... , \Delta f_{r-1}^k)$ by modifying the value of t. This operation can be made in some dynamic way, because each solution to [PLP] with a given t is obtained very quickly. Using some computer graphics, this enables us to develop an effective man-machine interface as a decision support system. This idea was originally realized by Korhonen-Wallenius (1987) in a slightly different way.

Remark 4.1   The reason why we take f_r, which may be relaxed most, as the objective function in [PLP] is as follows: Since the set $f(X)+R_+^r$ is convex in convex programming such as multi-objective linear programming, the new aspiration point based on (4.1), which is on the supporting hyperplane for $f(X)+R_+^r$, is not always much enough to compensate for the improvement of f_i ($i \in I_I$). Therefore, we apply the equal distribution (4.1) for objective functions to be relaxed except for f_r, and decide the amount of relaxation of f_r by solving [PLP], because this solution gives usually more relaxation than the one by (4.1).

Now we shall show some theoretical background to the above procedure (see Nakayama 1990, for details).

Lemma 4.1
    Let (x^k, z^k) be the solution to [ALP]. Then $(x^k, 0)$ is a solution to [PLP] with t=0.

Next, we shall show that we can obtain a solution to [PLP] with t=0 in one pivotting for the final tableau of [ALP] in many cases. To this end, the following auxiliary LP plays an important role.

[ALP']   Minimize   z

subject to $c_i x - f_i^k \leq (1/w_i)z$, $i=1,\ldots,r$

$x \in X$,

where $X := \{x \in R^n | \ Dx \leq d, \ x \geq 0\}$.

In order to derive the result to be shown, we use the tableau formula. Introducing slack vectors $u \geq 0$ and $v \geq 0$, we transform inequality constraints of [ALP] into the following equality constraints:

$$c_i x - (1/w_i)z^+ + (1/w_i)z^- + u \ = \bar{f}_i^k, \quad i=1,\ldots,r \ \left.\begin{array}{c} \\ \\ \end{array}\right\}$$
$$Dx \qquad\qquad\qquad\qquad + v = d \qquad\qquad \qquad (4.2)$$

A similar formula is available for [ALP'] by replacing $\bar{f}_i^k$ with f_i^k. Let C denote the matrix whose (i,j)-element is c_{ij}, and denote the coefficient matrix of (4.2) by

$$
A = \begin{pmatrix}
 & & -1/w_1 & 1/w_1 & 1 & & & \\
 & C & \vdots & \vdots & & \diagdown & & \\
 & & \vdots & \vdots & & & \diagdown & \\
 & & -1/w_r & 1/w_r & & 1 & & \\
 & & 0 & 0 & & & 1 & \\
 & D & \vdots & \vdots & & & & \diagdown \\
 & & \vdots & \vdots & & & & \diagdown \\
 & & 0 & 0 & & & & 1
\end{pmatrix}
$$

In the following, we consider [ALP], [ALP'] and [PLP] in a form with reformulated equality constraints like (4.2).

Lemma 4.2

The optimal basic matrix of [ALP] is also an optimal basic matrix to [ALP'].

Lemma 4.3 Exactly one of z^+ and z^- is basic in [ALP'] and [PLP].

From the above, we can suppose that z^- is basic in [ALP'] without loss of generality. In addition, let a_z- be the s-th column vector of the optimal basic matrix B in [ALP']. Note that [PLP] with t=0 is given by changing the coefficient vector $a_z+=(-1/w_1,\ldots,-1/w_r,0,\ldots,0)$ in A associated with z^+ in [ALP'] into $a'_z+=(0,\ldots,0,-1/w_r,0,\ldots,0)$ and by a similar change for a_z-. Let B' be a matrix obtained by replacing s-th column vector of B with

$$a'_z-=(0,\ldots0,1/w_r,0\ldots,0).$$

The following lemma shows that $(B')^{-1}$ is the inverse of an optimal basic matrix in [PLP] with t=0 and can be obtained in one pivotting from B^{-1} in many cases.

Lemma 4.4

If b_{sr}, the (s,r) component of B^{-1} is nonzero, the inverse of the optimal basic matrix of [PLP] with t=0 can be obtained in one pivotting for that of [ALP].

Remark 4.2

If $b_{sr}=0$, the inverse of the optimal basic matrix of [PLP] with t=0 can be obtained in a few pivotting for that of [ALP]. To see this, note that

$$B^{-1} B' = (e_1, \ldots, e_{s-1}, b'_s, e_{s+1}, \ldots, e_q)$$

where $e_i = (0, \ldots, 0, 1, 0, \ldots, 0)$ whose i-th component is 1. It follows then that if the s-th component of b'_s is 0, the matrix B' is singular because B^{-1} is nonsingular. This implies that B' obtained by replacing the s-th column vector of B with $a'_z -$ can not be basic in [PLP] with t=0. Since either z^- or z^+ is necessarily basic in [PLP] with t=0 according to lemma 4.3, some of other column vectors of B than $a_z -$ should also be replaced with appropriate vectors so that we might get an optimal basic matrix in [PLP] with t=0. In the final tableau in [ALP'], find an element $b_{sj} \neq 0$ where the j-th simplex criterion $p_j = 0$. It is assured that this b_{sj} does exist. For, if there is no p_j with the value 0, then we would have no optimal basic vectors other than B which leads to a contradiction in our situation. In a similar fashion, it leads to a contradiction to suppose that there is no $b_{sj} \neq 0$ for which the j-th simplex criterion is 0. Now take the j-th vector a_j as a basic vector. Namely, sweep out the j-th column vector by the pivot b_{ij}, where the subscript i is decided in the same way as in the usual simplex method, i.e.,

$$\theta_i = \min_k \{ b_{k,0}/b_{k,j} \mid b_{k,j} > 0 \}$$

If we have $b_{sr} \neq 0$ by the above pivotting, then we can apply Lemma 4.3 for the updated basic matrix B. Otherwise we continue the same proceddure until we attain this situation. This situation can be attained necessarily in finite steps. For, unless we get $b_{sr} \neq 0$ for ever, $a'_z -$ can not be basic, as stated above.

Remark 4.3

In cases where we want to make some of constraints, say f_i, to be objective functions, we can make a similar procedure as above by changing the coefficient of z in the equation corresponding to f_i in [PLP] from 0 into $1/w_i$.

As stated above, we can get the exact trade-off information very rapidly by using parametric optimization techniques in LP. Since decision makers are required to answer only their new aspiration levels of f_i to be improved, the burden of decision makers is very light and thus the interaction between decision makers and computers are carried very smoothly even though the problem has very many objective functions. Moreover, if decision makers are not satisfied with the obtained solution for the current aspiration, they can see the trade-off among objective functions along the direction of the present Δf_i (i I_I) and Δf_j (j I_P) very easily by modifying the value of t. Since the usual parametric optimization techniques can perform this very rapidly, if we use some graphic presentation for computer output, we can see the trade-off in a dynamic way, e.g. animation. This idea was originally suggested by Korhonen-Wallenius (1987). Such a device of man-computer interface is very important for making decision support systems.

5. Applications

5.1 Erection Management of Cable Stayed Bridge (Ishido et al. 1987)

In erection of cable stayed bridge, the following criteria are considered for accuracy control:
(i)   residual error in each cable tension,
(ii)   residual error in camber at each node,
(iii) amount of shim adjustment for each cable,
(iv)   number of cables to be adjusted.

Since the changge of cable rigidity is small enough to be neglected with respect to shim adjustment, both the residual error in each cable tension and that in each camber are linear functions of amount of shim adjustment.

Let us define n as the number of cable in use, ΔT_i (i=1,...,n) as the difference between the designed tension values and the measured ones, and x_{ik} as the tension change of i-th cable caused from the change of the k-th cable length by a unit. The residual error in cable tension caused by the shim adjustment is given by

$$p_i = |\Delta T_i - \sum x_{ik} \cdot \Delta l_k| \qquad (i=1,...,n)$$

We also define m as the number of nodes, Δz_j (j=1,...,m) as the difference between the designed camber values and the measured ones, and y_{jk} as the camber change of j-th node caused from the change of the k-th cable length by a unit. Then the residual error in the camber caused by the shim adjustments of $\Delta l_1,...,\Delta l_n$ is written by

$$q_i = |\Delta Z_i - \sum y_{jk} \cdot \Delta l_k| \qquad (i=1,...,n)$$

In addition, the amount of shim adjustment can be treated as objective functions of

$$r_i = |\Delta 1_i| \qquad\qquad (i=1,\ldots,n)$$

And the upper and lower bounds of shim adjustment inherent in the structure of the cable anchorage are as follows;

$$\Delta 1_{Li} \leq \Delta 1_i \leq \Delta 1_{Ui} \qquad\qquad (i=1,\ldots,n).$$

Fig. 5.1 shows one phase of erection management system of cable stayed bridge using the satisficing trade-off method. The residual error of each criterion and the amount of shim adjustment are represented by graphs. The aspiration level is inputted by the mouse on the graph. After solving the auxiliary min-max problem, the Pareto solution according to the aspiration level is represented by a graph in a similar fashion. This procedure is continued until the designer can obtain an appropriate shim-adjustment. This operation is very easy for the designer, and the visual information of trade-off among criteria is user-friendly. In addition, the graph of the Pareto solution can be animated according to various step sizes along the direction of the given trade-off by using the technique of the exact trade-off based on parametric optimization. This animated information of trade-off makes the designer to obtain his desirable solution more easily.

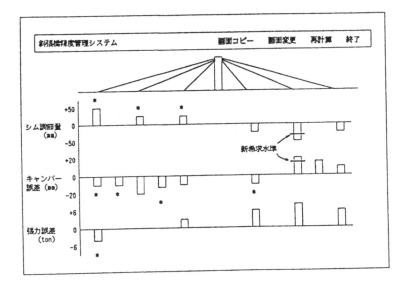

Fig. 5.1 A cable erection management system

5.2 Material Blending in Cement Production (Nakayama 1988)

Cement is produced by blending, crushing and burning several raw material stones such as lime, clay, silica, iron and so on. The provided raw material stones change from time to time. In order to keep the quality of cement at the desired level, the supply of each raw material stone is controlled depending on the change of chemical ingredient of raw material stones. Each factory usually produces several kinds of cement, and therefore the desired property of cement is various. It is needed to develop an effective method which leads to an appropriate solution for blending associated with the change of the kind of cement and that of ingredient of raw material stones. Usually, this material blending is implemented by a minicomputer linked with ingredient analyzers. To this aim, the satisficing trade-off method seem to work well. Some results of our experiments will be shown in the following.

	CaO	SiO_2	Al_2O_3	Fe_2O_3
lime	50.8	5.1	1.1	0.6
clay	4.0	63.6	15.7	7.2
silica	1.1	85.7	6.7	2.7
iron	1.5	27.4	3.3	67.8
another	12.9	41.5	15.1	7.7

Table 1.   An example of ingredient of raw material stones

The table 1 represents one of examples for chemical ingredients of each raw material stones. The criteria to be considered usually are the hydraulic modulus, the silica modulus and the iron modulus. Let x_i denote the amount of the i-stone to be used, and let C_i, S_i, A_i and F_i respectively denote the amount of CaO, SiO_2, Al_2O_3 and Fe_2O_3 contained in the i-stone. Each criterion is given by

i) hydraulic modulus: $HM = \sum_{i=1}^{5} C_i x_i / \sum_{i=1}^{5} (S_i + A_i + F_i) x_i$

ii) silica modulus: $SM = \sum_{i=1}^{5} S_i x_i / \sum_{i=1}^{5} (A_i + F_i) x_i$

iii) iron modulus: $IM = \sum_{i=1}^{5} A_i x_i / \sum_{i=1}^{5} F_i x_i$

The following constraints are imposed:

1) total amount to be blended:

$$\sum_{i=1}^{5} x_i = x_0$$

2) limitation of material supplier:

$$L_i \leq x_i \leq U_i \quad (i=1,\ldots,5)$$

For material blending problems in cement production, some of objective functions are of linear fractional form as stated above. Therefore the auxiliary min-max problem (Q) becomes a linear fractional programming, and hence we need some device for solving it.

An Algorithm for Linear Fractional Min -Max Problem

Let each objective function be of the form $F_i = p_i(x)/q_i(x)$ $(i=1,\ldots,r)$ where p_i and q_i are linear with respect to x. Then since

$$F_i(x) - F_i^* = \frac{p_i(x) - F_i^* q_i(x)}{q_i(x)} := \frac{f_i(x)}{g_i(x)},$$

the auxiliary Min-Max problem (Q) becomes a kind of linear fractional Min-Max problem. For this kind of problem, several methods have been developed: Here we shall use a Dinkelbach type algorithm (Borde-Crouzeix 1987, Ferland-Potvin 1985) as is stated in the following:

Step 1: Let $x^0 \varepsilon X$. Set $\theta^0 = \underset{1 \leq i \leq r}{Max} \; f_i(x^0)/g_i(x^0)$ and k=0.

<u>Step 2</u>:   Solve the problem

(P_k) $T_k(\theta^k) = \underset{x \varepsilon X}{\text{Min}} \quad \underset{1 \leq i \leq r}{\text{Max}} (f_i(x) - \theta^k g_i(x))/g_i(x^k)$

Let x^{k+1} be a solution to (P^k).

<u>Step 3</u>:   If $T_k(\theta^k) = 0$ then stop: θ^k is the optimal value of the given Min-Max Problem, and x^{k+1} is the optimal solution.

<u>Step 4</u>:   If $T_k(\theta^k) \neq 0$, take $\theta^{k+1} = \underset{1 \leq i \leq r}{\text{Max}} \quad f_i(x^{k+1})/g_i(x^{k+1})$. Replace k by k+1 and go to Step 2.

Note that the problem (P_k) is the usual linear Min-Max problem. Therefore, we can obtain its solution by solving the following equivalent problem in a usual manner:

(Q_k) Minimize z

subject to $(f_i(x) - \theta^k g_i(x))/g_i(x^k) \leq z$, $i = 1, \ldots, r$

$x \varepsilon X$

For material blending in cement production, many objective functions are to be attained at the desired levels. Since the desired levels are flexible to some extent, the decision maker is asked to answer his desirable level with its allowable tolerance range for each criterion. After this, each criterion is treated as two objective functions, for example,

$$HM \geq \overline{HM} - \varepsilon, \qquad HM \leq \overline{HM} + \varepsilon$$

where $\overline{HM}$ and ε denote the desirable level (target) and the allowable range for the hydraulic modulus, respectively.

As an example, we will show a result for the data in Table 1. At present, many factories operate the blending by use of the goal programming in which the above three criteria are considered (Kito et al. 1978). For the normal cement, the desired level of each criterion is $(HM, SM, IM) = (2.08, 2.59, 1.83)$.

In the goal programming, the deviation of each criterion from its target (originally it is of linear fractional form) is transformed into some linear objective function so that the simplex code for LP may be applied. Due to this, the weighting for each objective function is very difficult, because people can not know the quantative relation between the weight and the corresponding solution. Even if the decision maker wants to improve some of criteria and increase the corresponding weight, the obtained solution often causes other criteria worse too much. Many trial-and-errors are usually needed in order to get an appropriate weight.

By using the satisficing trade-off method, we can avoid this trouble. Since the obtained solution by the satisficing trade-off method has 'equality' to attain the target for each criterion, the setting or change of the target and allowable range for each criterion is easy.

Case 1:
For our problem, we set the allowable error to be 0. Then we obtain the solution which attains the exactly desired level of each criterion.

	Pareto sol.	Target	Allowable Range
HM	2.080	2.080	0.0
SM	2.590	2.590	0.0
IM	1.830	1.830	0.0

$x_1 = 573.713$, $x_2 = 26.458$, $x_3 = 23.136$, $x_4 = 1.693$, $x_5 = 95.000$

Case 2:
In many factories, the cost for blending is not taken into account up to now. We shall show results for cases in which the cost is taken into account:

$$\text{iv) cost:} \quad P = \sum_{i=1}^{5} p_i x_i$$

For the same aspiration level and the allowable error as the above, we set the aspiration level of cost to be 15.0 with the allowable range 3.0. The result is given by the following:

	Pareto sol.	Asp.Level		Lowest	Highest
		(Target	Range)		
F1 (tar)	2.1839	2.0800	.0000	.8706	3.1961
F2 (tar)	2.6968	2.5900	.0000	1.3472	5.8837
F3 (tar)	1.8017	1.8300	.0000	.3894	2.1598
F4 (min)	15.6700	15.0000		7.2000	52.9500

$x_1 = 599.8356$, $x_2 = 100.0000$, $x_3 = 3.5264$, $x_4 = 2.5906$, $x_5 = 14.0474$

If the decision maker agree with the increase the aspiration level of cost to 25.0, we can get the following solution by the satisficing trade-off method.

	Pareto sol.	Asp.Level		Lowest	Highest
		(Target	Range)		
F1 (tar)	2.0800	2.0800	.0000	.8706	3.1961
F2 (tar)	2.5900	2.5900	.0000	1.3472	5.8837
F3 (tar)	1.8300	1.8300	.0000	.3894	2.1598
F4 (min)	22.3000	25.0000		7.2000	52.9500

$x_1 = 588.8332$, $x_2 = 100.0000$, $x_3 = 0.4293$, $x_4 = 2.3431$, $x_5 = 28.3944$

This solution shows that it costs at least 22.3 to attain the requirement aspiration level. Recall that the cost for the solution of the case 1) is 56.662. It can be seen that taking the cost into account, we can get a solution with the cost less than half of case 1), while other criteria attain at the desirable levels.

Case 3:
    If the criteria are flexible to some extent around the target, we can get another solution with less cost. The following is the result for the case with the allowable range 0.1 around the target for each criterion.

	Pareto sol.	Asp.Level		Lowest	Highest
		(Target	Range)		
F1 (tar)	2.1794	2.0800	.1000	.8706	3.1961
F2 (tar)	2.6894	2.5900	.1000	1.3472	5.8837
F3 (tar)	1.7302	1.8300	.1000	.3894	2.1598
F4 (min)	14.9963	15.0000		7.2000	52.9500

$x_1 = 600.3790, \quad x_2 = 100.0000, \quad x_3 = 4.5450, \quad x_4 = 3.4324, \quad x_5 = 11.6436$

It is seen that we can get a solution with less than one third cost of the case 1), while other criteria remain within an allowable error. As can be seen through the above experiments, we can easily obtain a solution as we desire from a total viewpoint. Unlike the traditional goal programming, since we search the solution by adjusting the aspiration level, which is very easy to answer, and has a direct quantitative relation with the corresponding Pareto solution, the satisficing trade-off method seems very operational. The method is working in the material blending system of some cement company in Japan.

6. Concluding Remarks

In this paper, the aspiration level approach to multi-objective programming is introduced along with some device for exact trade-off using the parametric optimization. In some kind of practical problems such as the long term planning, the visual (graphical) information of trade-off is very effective, because the precise numerical value of criteria is not necessarily needed, but the rough tendency is enough. In addition, the animated information such as Pareto race (Korhonen-Wallenius 1988) and the exact trade-off method using the parametric optimization (Nakayama 1990) is more user-friendly and more easy to handle.

On the other hand, in another kind of problems such as the material blending in cement production and the bond portfolio, decision makers are accustomed with numerical values of criteria. Therefore, in such problems, we do not need to use graphical information. It seems important to make an appropriate man-machine interface according to the problem.

REFERENCES

Borde, J. and Crouzeix, J.P. (1987), Convergence of a Dinkelbach-type Algorithm in Generalized Fractional Programming, Zeitschrift fur Operations Research 31, 31-54.

Ferland, A. and Potvin, J. (1985), Generalized Fractional Programming: Algorithms and Numerical Experimentation, European J. Operational Research 20, 92-101.

Grauer, M., Lewandowski, A. and Wierzbicki A.P. (1984), DIDASS Theory, Implementation and Experiences, in M. Grauer and A.P. Wierzbicki (eds.) Interactive Decision Analysis, Proceedings of an International Workshop on Interactive Decision Analysis and Interpretative Computer Intelligence, Springer: 22-30.

Ishido, K., Nakayama, H., Furukawa, K., Inoue, K. and Tanikawa, K. (1987). Management of Erection for Cable Stayed Bridge using Satisficing Trade-off method, in Y. Sawaragi, K. Inoue and H. Nakayama (eds.) Toward Interactive and Intelligent Decision Support Systems, Springer: 304-312.

Kallio, M., Lewandowski, A. and Orchard-Hays, W. (1980), An Implementation of the Reference Point Approach of Multiobjective Optimization, WP-80-35, IIASA

Kito, N. and Misumi, M. (1978), An Application of Goal Programming to Cement Production, Communications of the Operations Research Society of Japan 23, 177-181.

Korhonen, P. and Wallenius, J. (1987). A Pareto Race, Working Paper F-180, Helsinki School of Economics, also Naval Research Logistics (1988)

Korhonen, P. and Laakso, J.: Solving Generalized Goal Programming Problems using a Visual Interactive Approach, European J. of Opertional Res., 26, 355-363 (1986)

Nakayama, H. (1984), Proposal of Satisficing Trade-off Method for Multi-objective Programming, Transact. SICE, 20: 29-35 (in Japanese)

Nakayama, H. and Sawaragi, Y. (1984). Satisficing Trade-off Method for Interactive Multiobjective Programming Methods, in M. Grauer and A.P. Wierzbicki (eds.) Interactive Decision Analysis, Proceedings of an International Workshop on Interactive Decision Analysis and Interpretative Computer Intelligence, Springer: 113-122.

Nakayama, H. (1985), On the Components in Interactive Multiobjective Programming Methods, in M. Grauer, M. Thompson and A.P. Wierzbicki (eds.) Plural Rationality and Interactive Decision Processes, Springer: 234-247.

Nakayama, H. and Furukawa, K. (1985), Satisficing Trade-off Method with an

Application to Multiobjective Structural Design, Large Scale Systems
8: 47-57.

Nakayama, H., Nomura, J., Sawada, K. and Nakajima, R. (1986). An Application of Satisficing Trade-off Method to a Blending Problem of Industrial Materials, in G. Fandel et al. (Eds.), Large Scale Modelling and Interactive Decision Analysis. Springer, 303-313.

Nakayama, H. (1988), Satisficing Trade-off Method for Problems with Multiple Linear Fractional Objectives and its Applications, presented at the International Conference on Multiobjective Problems in Mathematical Programming, Yalta/USSR, October 1988.

Nakayama, H. (1989), An Interactive Support System for Bond Trading, in A. G. Lockett and G. Islei (eds.) Improving Decision Making in Organizations, Springer, 325-333

Nakayama, H. (1990), Trade-off Analysis using Parametric Optimization Techniques, Research Report 90-1, Dept. Appl. Math., Konan University

Sawaragi, Y., Nakayama, H. and Tanino, T. (1985). Theory of Multiobjective Optimization, Academic Press

Steuer, R. E. (1986). Multiple Criteria Optimization: Theory, Computation, and Application, Wiley

Ueno, N., Nakagawa, Y., Tokuyama, H., Nakayama, H. and Tamura, H., A Multiobjective Planning for String Selection in Steel Manufactureing, submitted for publication

Wierzbicki, A. P. (1981), A Mathematical Basis for Satisficing Decision Making, in J. Morse (ed.), Organizations: Multiple Agents with Multiple Criteria, Springer: 465-485.

Wierzbicki, A. (1986), On the Completeness and Constructiveness of Parametic, Characterization to Vector Optimization Problems. OR spectrum, 8: 73-87.

A HIERARCHICAL APPROACH TO PERIODIC SCHEDULING OF LARGE SCALE TRAFFIC LIGHT SYSTEMS

C. Pascolo
C.O.R.E.L. Friuli, Udine, Italy

P. Serafini
Università di Udine, Udine, Italy

W. Ukovich
Università di Trieste, Trieste, Italy

ABSTRACT: A method for the fixed-time signal traffic light control problem is proposed for large metropolitan areas. Its characteristics are: a multiobjective formulation focusing on desired offsets, a hierarchical structure derived from the problem in order to reduce complexity and an interactive strategy for implicitly assessing a performance measure. The core of the method is a mathematical model for scheduling periodic phenomena.

1. INTRODUCTION

This paper deals with the well-known problem of designing a fixed-time signal control. Loosely speaking, an appropriate periodic schedule is sought for the signal lights controlling a given area in order to produce suitable traffic conditions.

This problem has been extensively investigated in the past decades under different points of view and with different methods, and many important results have been obtained both of theoretical and of practical relevance (among the many others, see for instance [1], [2], [3], [4]).

Nevertheless, the research in this area is still in progress; in fact, the continuing evolution of the control device technology and design methods not only opens new directions towards radically innovative approaches to deal with traffic control (as for instance in [5], [6], [7], [8], [9]), but also stimulates more advanced and effective solutions for classical problems (see for example [10], [11]).

This paper adopts the latter attitude, aiming at new formulations and solution methods for a large scale version of an old problem like the fixed-time signal control design by extending the approach presented in [12].

The difficulty of (and the interest for) this problem originates not only from its dimensions, as measured for instance by the number of vehicular flows to be controlled, but also from the complexity of the relations and interactions between them. Rather than in the regularly meshed American structures, such features may be much more relevant in the old and squeezed European cities, where different traffic conditions and flow characteristics often coexist (not always peacefully), and only scarce space resources are available to accomodate all traffic flows.

In such severe conditions it may be difficult even to provide a satisfactory formalization of the problem itself; in particular, if a classical optimization approach is adopted, choosing meaningful and effective criteria for the system performance evaluation may constitute a very critical problem in its own right.

To tackle such difficulties, the approach proposed in this paper has the following features:

– it is a multiobjective optimization approach: the different factors affecting the system performance (typically, capacity, queueing delays and offsets) will be considered separately, without pretending to aggregate them a priory into a single objective;

– it has a hierarchical structure: as it will be discussed below, the problem itself has a hierarchical nature, that it is convenient to exploit in order to cope with the problem large scale;

– it operates in an interactive way: the design procedure we propose allows (but not necessarily requires) an active intervention of the designer; in this way all the designer's experience and knowledge of the problem under interest may be effectively exploited without requiring a pedantic formulation of all the problem details within preassigned schemes;

– it relies on a compact mathematical model which enters all hierarchy levels with the same structure;

– it relies on an advanced graphical data base and visual system to enhance the interactive capabilities of the traffic engineer.

As a result, we propose a Decision Support System for solving the fixed-time control problem. The DSS is designed so as to be a versatile framework into which hardly formalizable or quantifiable features may be effectively taken into account, yet providing precise and sound procedures to deal with.

In Section 2 the problem is formulated in terms of design variables, rigid problem constraints and "goal" constraints, which consist in desired values for green durations and offsets. In Section 3 the hierarchical structure is discussed and a procedure to deal with it is proposed on the basis of a mathematical model which is investigated in Section 4. The interactive part of the method is presented in Section 5 and the graphical system is briefly described in Section 6. Finally, some conclusions are drawn in Section 7.

2. PROBLEM FORMULATION

The problem considered in this paper consists in designing a periodic fixed-time signal plan for a complex system of intersections in a large metropolitan area. The paper mainly focuses on large scale problems of this kind, which either the dimensions or the complexity make especially awkward to tackle with the usually available methods.

In this section we present a description of the problem we are dealing with. Although the formulation we consider complies in several aspects with the usual way this problem is approached to in the literature and in the professional practice, there are some original features worth of consideration.

A signal plan is determined by a cycle time T, i.e. the period, and by the time instants within the cycle when each light of each signal has to be turned on and off. Using effective rather than actual timings allows to deal only with green lights (effective reds being just their complement with respect to the cycle). So our problem consists in choosing a cycle time and in placing green starts and ends within the cycle.

In our formulation each signal that may be driven independently will be considered separately: no a priori fixed temporal relation is required nor supposed among signal greens, such as for instance a given phase sequence for multiphase intersections.

So the problem variables are the times, within the cycle, at which each independent signal has to be turned on and off and the cycle time T. The problem variables are subject to a number of constraints, expressing the conditions the designer requires for the traffic flows under control.

First we consider constraints which must be obeyed in any case by the traffic flows. These are referred to as *rigid* constraints and consist of:

– minimum clearance times between greens of conflicting signals;

– rigid coordinations between the greens of strongly interacting signals controlling the same flow (for instance, in the case when a queue at a downstream signal may produce jamming by blocking lateral flows).

Besides the above rigid constraints, we also consider another class of requirements. Let us first take into account the capacity of the intersection. As a preliminary requirement a nominal lower bound should be set for each green duration to allow all traffic flow within the cycle to go through the corresponding stopline. However, we may also wish to find a signal plan with all lower bounds on green durations larger than the nominal values by a common factor $\mu > 1$, which can be regarded as the reserve capacity of the intersection (see for instance [13,14]). So we are interested in the largest possible value for μ for which a signal plan exists. On the other hand there could be situations for which only overloaded solutions ($\mu < 1$) exist and we are forced to accept this bad performance. Note that the maximum reserve capacity is an increasing function of the green durations and the cycle time.

Secondly we may consider queueing delays at the stoplines. In the literature a performance evaluation function of the queueing delay withn respect to the green duration and the cycle time is considered which is convex in the green duration ([13]). Loosely speaking this function is decreasing with respect to the greeen durations and increasing with respect to the cycle time. So these two performance evaluation criteria are in conflict for what concerns the cycle time. These constraints involving the lower bounds on green durations are called *hard goal* constraints.

A third criterion involves signals whose interaction is not so strong to justify rigid coordination conditions, yet sensible enough to influence quantities affecting the system performance, such as stops, waiting and throughput times, etc. For such signals we assume that "desired" intervals are provided within which the relative offsets should lie in order to guarantee a satisfactory level for the system performance. Then we look for a feasible solution with offsets possibly within, or at least as close as possible to, the given desired intervals. This third type of constraints is called *soft goal* constraints. Both hard and soft goal constraints are regarded as *flexible* constraints in the sense that they can be revised if necessary.

Hence we model the problem by turning objective functions into flexible constraints which guarantee that some goals can be met if a feasible solution does exist. More exactly objectives regarding capacity and queueing delays are reflected into appropriate lower bounds for the green durations, and offsets requirements are converted into coordinations. It is important to point out the this is possible and convenient exactly because rigid and flexible constraints have the same mathematical structure. The problem of finding a signal plan meeting such conditions is the central subject of [12], which the reader is referred to for a comprehensive presentation. For the purpose of this paper, however, the concepts just sketched should suffice for a thorough understanding.

There are two alternative approaches to find efficient feasible solutions. By "efficient" we mean a solution which cannot be made better off with respect to any objective without worsening any other one. In the first approach we set the goals at high levels. A solution with all objectives meeting these goals corresponds to the so-called "utopia" point, and it is expected not to be available in general. In this case we must relax some goals until a feasible solution can be found, which turns out to be efficient.

Note that with this approach the different objectives are not aggregated a priori into a single performance index; rather, an evaluation criterion is implicitly assessed through an interactive procedure with the traffic designer, who uses its experience and knowledge of the process to relax the flexible constraints. In this sense the procedure we propose is similar to the interactive approaches of multi-criteria decision making procedures (cf. for instance [15]).

The goal relaxing procedure could also be performed in an automatic way, according to some predetermined evaluation method. So the approach we propose to design a signal plan iterates two tasks:

- trying to compute a feasible solution;
- relaxing the goal constraints.

In the second approach the traffic engineer sets the goals to rather low values so that a feasible solution is expected to exist. The algorithm we have developed is able to provide all feasible solutions which are different in their combinatorial structure. If all these solutions are not very many, as it typically happens, we may optimize each one of them in traditional way with respect to the stated objective functions.

We may also exploit this strategy in order to set the cycle time. Indeed, once the combinatorial structure of a solution is fixed, the determination of the minimum cycle time such that a signal plan exists can be easily carried out through network flow techniques. Alternatively we could also compute the minimum cycle time by using a binary search technique without computing preliminarly the combinatorially different solutions.

Both approaches are made available to the traffic engineer which can switch between them according to his experience and perception of the problem at hand.

3. A HIERARCHICAL APPROACH

In principle, the two tasks introduced in the last section, performing the feasibility and the relaxation actions, could act on a global scale, according to [12]. The algorithm presented in [12] performs well on real problems of size up to about thirty signals, which roughly correspond to very complex intersections within a limited area. However, a similar nice computational behaviour could hardly be expected for larger systems (say of over one hundred signals involving several intersections) due to the fact that the problem is NP-complete.

To overcome such a difficulty we propose to exploit the intrinsic hierarchical nature of the problem in order to decompose it in a sequence of simpler subproblems.

At the first level only rigid and hard goal constraints appear. It is convenient to think of the signals as nodes of a graph $G1$, with arcs corresponding to constraints between two signals. Since offsets are not considered at this level, distant traffic lights are not constrained each other and $G1$ is likely to be a disconnected graph. We shall refer to each of its connected components as an "intersection", according to the common notion of an intersection as a limited area within which several traffic flows compete for using a common space resource.

The feasibility problem at this level is so split into several independent subproblems, each correspondig to a connected component, whose feasible solutions give a complete schedule for all signals, i.e. the timings of the two events corresponding to turning each signal to green and to red. Note that the feasibility of such timing values is not affected by adding the same constant (modulo the cycle) to all the signals of a same intersection.

As it will be shown below, if no feasible solution exists for a subproblem, the method produces a circuit whose nodes cannot be assigned timings consistent with all the arc constraints involved. Therefore some flexible constraints must be appropriately relaxed in order to allow a feasible solution to exist. Discussing how to do this job is deferred to Section 5, after discussing the algorithm details.

The second level considers the objectives corresponding to the most stringent offsets between signals of different intersections. A value for each of the above mentioned additive constants has then to be found for the goals relative to these offsets to be satisfied.

This is accomplished as follows: let t^i be the constant to be determined for intersection i and τ_k^i be the timing of the k-th event (the turning to green or to red of some signal) of the same intersection. Let d^- and d^+ be the lower and upper extremes of the desired interval for the offset between the k-th event of intersection i and the h-th event of intersection j (we dispense with the obvious appropriate indexing). Then we must have

$$d^- \leq t^j + \tau_h^j - t^i - \tau_k^i + zT \leq d^+,$$

for some integer z, that is

$$d^- + \tau_k^i - \tau_h^j \leq t^j - t^i + zT \leq d^+ + \tau_k^i - \tau_h^j.$$

This represents a constraint for the pair t^i, t^j. Since at this level t^i is the only variable associated to each intersection, it is convenient to transform $G1$ by collapsing all nodes corresponding to the signals of a same intersection into a unique node representing the whole intersection. A new graph $G2$ is thus produced with arcs corresponding to the offsets considered at this level.

$G2$ may consist of several connected components which we call clusters of intersections. For computational reasons, all offsets within a same cluster must be taken into consideration at this level.

Thus the second level problem consists in timing the nodes of $G2$ according to the constraints displayed above. Again, it splits in separate subproblems, one for each cluster.

A method to solve them will be presented in the next section. Also in the second level infeasibility results are handled by relaxing offset constraints. As for $G1$, the feasibility of the timings found for $G2$ is not affected by adding the same constant (modulo the cycle) to all the signals of a same cluster.

Now the clusters of the second level may be correlated by introducing higher level offsets for pairs of signals lying within different clusters. Exactly the same situation is thus produced as for the previous level and again the same procedure is applied to graph $G3$, obtained from $G2$ by collapsing each cluster into a single node.

The whole process of creating higher levels may then be iterated as far as convenient, until either a connected graph is obtained or all requirements have been considered.

Deciding how many levels to generate and which offsets to put into each of them is largely a matter of choice open to the designer's attitude. It must only pointed out that all offsets among signals lying into two different connected components must be considered at the same hierarchical level and that all requirements of each level must be fulfilled (though possibly in a relaxed version) before accessing to an upper level.

In fact, the method we propose does not consider the possibility of backtracking to and solving again lower levels in order to fulfill the upper level goals. Indeed, allowing such a possibility would be equivalent to solving the problem on a global scale with the non hierarchical approach presented in [12].

Besides these considerations, a tradeoff between the problem size and the available computational resources is likely to suggest sound criteria for the most appropriate number and size of the levels, whereas the specific features of the problem to be solved should naturally provide appropriate guidelines for the composition of each level.

4. THE BASIC MATHEMATICAL MODEL

Let T be a positive real number (the cycle length) and d_{ij}^-, d_{ij}^+ ($i, j = 1, \ldots, n$; $i \neq j$) be real numbers such that $d_{ij}^- \leq d_{ij}^+$ (the lower and upper bounds for the ordered pair i, j).

We are interested in finding real numbers t_i, $i = 1, \ldots, n$ such that

$$d_{ij}^- \leq t_j - t_i + z_{ij}T \leq d_{ij}^+ \tag{1}$$

for some integers z_{ij}.

A theoretical investigation of this problem is beyond the scope of this paper. It can be found in [16]; in [12] this problem is presented in a modified version to handle signal constraints. We just point out here that $d_{ij}^+ - d_{ij}^- \geq T$ corresponds to no constraint for the pair (i, j).

The above problem is called "Periodic Event Scheduling Problem" (PESP). It can be also modelled through a directed graph G with n nodes and m arcs (i, j) whenever $d_{ij}^+ - d_{ij}^- < T$. This problem can be easily shown to be NP-complete via a transformation from the Hamiltonian circuit problem.

The hardness of the problem stems from the fact that, besides the reals t_i, one has to find also the integers z_{ij}. For fixed z_{ij}'s the problem is polynomial and can be solved by exploiting the techniques developed in [17] which in turn are based on the Dijkstra's algorithm for finding shortest paths in a network. We note that the number of independent integer variables is actually equal to the number of independent circuits of the graph G.

In order to find the solution the algorithm guesses in an implicit enumeration fashion values z_{ij} until either a fortunate guess leads to a feasible solution or it can be proven that no feasible solution exists. The algorithm explores a search tree trying to find paths to the bottom leaves (if any).

An interesting feature of the algorithm is that to each leaf of the search tree which is not a bottom node a blocking circuit C can be associated. A blocking circuit is a circuit whose nodes cannot be given any value t_i without affecting feasibility of at least one arc in it, since

$$\Delta(C) := \sum_{(i,j)\in C^+} \left(d_{ij}^- - z_{ij}T\right) - \sum_{(i,j)\in C^-} \left(d_{ij}^+ - z_{ij}T\right) > 0$$

where C^+ is the set of arcs of the circuit C directed as the circuit, and C^- is the set of arcs directed opposite to the circuit C. The quantity $\Delta(C)$ is called the blocking gap, and it is provided too by the algorithm. Therefore when infeasibility is detected a set of blocking circuits is given by the algorithm.

It is clear that if we want to resume the algorithm from the point where a certain blocking circuit was found we must relax the constraints of the arcs of that blocking circuit by at least the blocking gap.

Relaxing the constraints in order to make PESP feasible can be done by simply relaxing arc constraints of blocking circuits. Since the algorithm provides several blocking circuits, we may exploit such a freedom of choice in order to relax the constraints at the highest possible system performance. We shall investigate this problem in more detail in the next section.

5. GOAL RELAXATION METHODS

We recall that goal constraints are represented by interval constraints (1) of the PESP model. Thus relaxing a goal consists in widening the corresponding interval given by (1). As previously explained, if the PESP algorithm returns an infeasibility result at some hierarchical level, it outputs also a number of blocking circuits. The first decision the traffic engineer has to face in order to reach fesibility concerns the choice of the blocking circuit to be relaxed. The following alternative strategies may be adopted: select the blocking circuit

– which was generated by the algorithm at the maximum search tree depth. This rule is motivated by computational reasons because it speeds up the search for a feasible solution;

– with the smallest blocking gap. This rule is motivated by system performance considerations. In fact, smallest blocking gaps mean smallest goal relaxation and guarantee Pareto efficiency within the hierarchical level;

– made up of the "more relaxable" arcs. In other words, the selection is left to the free designer's choice. This rule is motivated by the convenience of implicitly taking into account hardly formalizable performance requirements, and is therefore more suited to solicit designer's preferences.

Once a particular blocking circuit is selected, another decision has to be made about how much to relax each single arc of the circuit. Of course, the total relaxation along the circuit has to be at least equal to the blocking gap. Again, three strategies can be alternatively followed:

– select the last arc inserted by the algorithm. This choice is motivated by algorithmic efficiency since it enables the algorithm to go one step deeper in the search tree;

– relax all the arcs of the circuit in a uniform way. In this case the deviation from the utopia point is uniform for a subset of goals, and a better system performance with respect to the initial requirements can be expected;

– distribute the blocking gap according to the designer's free choice. Again, this corresponds to an implicit assessement of the system performance measure according to the designer's view.

Note that the first two strategies may be implemented automatically.

It is necessary to recall that the maximum amount a soft goal (i.e. an offset) can be relaxed corresponds to an interval of size T for that arc. Of course, this means no constraint at all in the mathematical model and the impossibility of imposing any satisfactory offset between certain pairs of signals in the real problem.

6. GRAPHIC SUPPORT

For the mathematical model to be a useful and efficient tool in the hands of the traffic designer a powerful graphic interactive system has to be built. The one we are proposing has the following features.

There is a graphical data base able to store a detailed map of the town. The user may retrieve any part of the map and watch it on the screen at the desired size. All data are input to the system through an interactive dialog with the map. Initially the user works separately for each intersection. Traffic lights are created by clicking the mouse on the corresponding position on the screen. After clicking, a numbered circle appears on the screen representing the traffic light.

Conflicts are created by clicking on the corresponding pairs of traffic lights and by a third click on the ideal intersection point of the two conflicting flows. The system computes automatically the distances of this point from both stoplines and consequently the clearance times according to given tables. These default values may be reset by the user if necessary.

Coordinations for a series of traffic lights traversed by a homogeneous flow are created by clicking in sequence on the corresponding traffic lights. The user has to provide the average value for the flow and its average velocity and the system computes default head and tail offsets. These values are displayed to the user who may reset them if necessary. Nominal values for the green lower bounds are also computed.

In our approach traffic flows may also intersect at some traffic light. In these cases the system creates "virtual" traffic lights each one of them taking care of one traffic flow. The virtual traffic lights are eventually merged into the actual traffic light. This feature is transparent to the user who need not even know the existence of virtual traffic lights.

Both coordinations and conflicts are displayed over the street map as arcs connecting the relative nodes, so that the intersection graph is immediately available to the user providing an indication of the complexity of the scheduling problem.

The user assesses a first value for the reserve capacity and the cycle time. Correspondingly the system computes the overall queueing delay and shows this value. Before passing the data to the algorithm the system performs some consistency checks.

At this point the user is free either to compute the minimum cycle time for the given reserve capacity or to compute a feasible signal plan for the given data. Computing the minimum cycle time is important in order to set a common cycle time for all intersections. Therefore this computation should be done first for all intersections in order to provide a common framework for the whole area.

Once a cycle time has been assessed feasible schedules are looked for as described in the previous sections. In case infeasibility is detected blocking circuits are stored by the system and the user may retrieve any one of them. A blocking circuit is displayed as a series of bar charts for the involved signals. The last bar corresponds again to the first signal of the blocking circuit and should be displaced by an integer number of periods with respect to the first bar in order to have feasibility. The constraints making this impossible (i.e. a too high lower bound for a green duration, or a too short head coordination, or a too long tail coordination, or a too short clearance time) are displayed as a series of segments connecting the corresponding events (green or red switching) on the bar chart diagram, thus making immediately clear to the user which constraints are involved and the amount of relaxation necessary to get rid of the blocking circuit. The user can relax the constraints by clicking on the relative event over the diagram and dragging it by the desired amount. During dragging the system shows the value of the performance indices given by the modified data, so that the user may appreciate directly the effect of the relaxation.

The user may pass to another blocking circuit if he is not satisfied of the relaxation needed. Eventually he resumes the algorithm and the procedure is iterated until feasibility is attained. Other feasible solutions may also be obtained by backtracking and restarting the algorithm from another blocking circuit. At the end the different solutions may be compared.

If on the contrary feasible solutions are available without relaxation the user may run an optimization problem of network flow type for each one of them by assessing relative weights between capacity and delays or alternatively, after seeing the value of the performance indices for the different solutions he may guess higher values for the goals and restart the algorithm.

After each intersection has been scheduled at the best possible performance index the user may connect different intersections in order to provide convenient offsets between signals of different intersections.

After the design process is terminated a simulation is also performed to better validate the signal plan. If bottlenecks are detected degrading the plan performance data can be possibly modified and the whole design procedure should be repeated.

7. CONCLUSIONS

A new approach has been presented to deal with the problem of designing a fixed-time signal plan for large scale intersection systems. It is based on a multi-objective hierarchical model which may be solved in an interactive way. Its overall philosophy and computational methods have been discussed.

The next step in this research will consist in applying the overall proposed approach to some real problems; only partial experiments have been carried out to date. For instance, the algorithm for solving the first level subproblem (see [12]) has been applied to several practical cases and the PESP routine has been tested for many randomly generated graphs. The computational evidence shows on both cases an excellent performance also in relatively large instances (up to about fifty nodes). Actually, the computational performance observed is far better than expected for such an NP-complete problem.

Obviously, an a posteriori evaluation of the characteristics of the method proposed in this paper is essential to draw conclusions about its effectiveness, especially because informal procedures and value judgements are involved. Nevertheless, an optimistic expectation may be based on the good outcomes of the available practical experience and on the intrinsic flexibility of the proposed approach.

Should these expectations be fulfilled, they would open a very stimulating way towards implementing these methods online. To such an extent, the goal relaxation task should better run automatically and the feasibility task could be advantaged by the fact that goal constraints are gradually updated, thus providing the possibility of using results of a previous computation (blocking circuits and gaps) to bypass several computational stages.

8. REFERENCES

[1] K. E. Stoffers, "Scheduling of traffic lights - a new approach", *Transportation Research*, 2, 199- 234, 1968.

[2] J. D. C. Little, "The synchronization of traffic signals by mixed-integer programmig", *Operations Research* 14, 568-94, 1966.

[3] D. I. Robertson, "TRANSYT: a traffic network study tool", RRL Report LR 253, 1969.

[4] N. H. Gartner, J. D. C. Little and H. Gabbay, "Optimization of traffic signal settings by mixed-integer linear programming", *Transportation Science*, 9, 321-63, 1975.

[5] J. D. C. Little, M. D. Kelson and N. H. Gartner, "MAXBAND: A Program for Setting Signals on Arteries and Triangular Networks", *Transportation Research Record*, N. 795, 40-6, 1981.

[6] M. Drouin, H. Abou-Kandil, G. Dib, P. Bertrand, "A new approach for real-time control of urban traffic networks", *Proc. IFAC/IFIP/IFORS 4th Int. Symp. on Traffic Control Systems*, Baden-Baden, 239-243, 1983.

[7] N. H. Gartner, "OPAC: A Demand Responsive Strategy for Traffic Signal Control", *Transportation Research Record*, 906, 75-81, 1983.

[8] R. Boettger, "Online Optimization of the Offset in Signalized Street Networks", *Proc. IEE Conference on Road Traffic Signalling*, London, 1982.

[9] A. J. Al-Khalili, "Urban Traffic Control - A General Approach", *IEEE Tr. on Systems, Man, and Cybernetics*, SMC-15, No. 2, 260-271, 1985.

[10] W. Dauscha, H. D. Modrow and A. Neumann, "On Cyclic Sequence Types For Constructing Cyclic Schedules", *Zeitschrift für Operations Research*, 29, 1-30, 1985.

[11] G. Improta and G. E. Cantarella, "Control System Design for an Individual Signalized Junction", *Transportation Research-B*, 18B, 147-67, 1984.

[12] P. Serafini, W. Ukovich, "A Mathematical Model for the Fixed-Time Traffic Control Problem", *European J. of Operational Research*, vol. 42 , p. 152-165, 1989

[13] B.G. Heydecker and I.W. Dudgeon, "Calculation of Signal Settings to Minimize Delay at a Junction", in *Transportation and Traffic Theory*, N.H. Gartner and N.H.M. Wilson (eds.), Elsevier, New York, 1987

[14] G.E. Cantarella and G. Improta, "Capacity Factor or Cycle Time Optimization for Signalized Junctions: a Graph Theory Approach", *Transportation Research B*, 22 (1) 1-23, 1988.

[15] C. L. Hwang and A. S. M. Masud, *Multiple Objective Decision Making - Methods and Applications*, Springer, Berlin, 1979.

[16] P. Serafini and W. Ukovich, "A mathematical model for periodic scheduling problems", *SIAM J. on Discrete Mathematics*, vol. 2, n. 4, p. 550-581., 1989

[17] R. T. Rockafellar, *Network Flows and Monotropic Optimization*, Wiley, 1984.

A MULTI-STAGE DECOMPOSITION APPROACH FOR A RESOURCE CONSTRAINED PROJECT SCHEDULING PROBLEM

P. Serafini, M.G. Speranza

Università di Udine, Udine, Italy

1. ABSTRACT

A Decision Support System for a resource constrained scheduling problem is described. Great attention is given to the interactive algorithmic support. The DSS is based upon a decomposition of the problem which adheres to the associated decision process. Combinatorial algorithms are introduced in order to support each of the resulting subproblems. The subproblems are described, together with the main features of the implemented prototype.

2. INTRODUCTION

A considerable research effort has been devoted to scheduling problems. The reason for such an interest is that they encompass many and very different problems, both in terms of characteristics, application fields and solution techniques. Mainly for historical reasons, a large class of scheduling problems have been dealt with separately, usually by means of network analysis and general optimization techniques (see Patterson, 1984, Christofides et al, 1987, and Speranza and Vercellis, 1990). On the other side theoretical results have been found for machine scheduling problems through combinatorial techniques leading to a detailed computational complexity analysis (see for instance Rinnooy Kan, 1976, and Lenstra and Rinnooy Kan, 1985). As a matter of fact, there is some overlapping between the two classes. For instance the problem of minimizing the total project duration under resource constraints is a generalization of the Job Shop Problem.

In this paper we deal with a problem proposed by Anthonisse et al. (1988) for an international exercise coordinated by the International Institute for Applied System Analysis, Vienna. This is a resource constrained project scheduling problem strongly related to the Job Shop Problem. Tasks have to be processed in order to minimize a weighted function of maximum earliness and tardiness and total earliness and tardiness. Constraints, which generalize classical precedence constraints, can be established between pairs of tasks. Renewable and discrete resources, available only in defined time intervals, are required in order to process each task. Each task can be processed in a number of modes, where each mode requires a subset of resources and a processing time. Moreover release dates and deadlines are associated to each task defining a time interval during which the task must be processed.

In this paper we address the problem of designing a Decision Support System for such a complex problem, so as to allow building effective solutions by means of a careful trade-off between algorithmic power and human interaction. After a description of the problem in Section 3 and a sketch of the development process of the DSS in Section 4, our decomposition approach to the problem is discussed in Section 5. The subproblems identified in the decomposition approach are analysed in Sections 6-8. In particular, the resource assignment problem is analysed in Section 6. The sequencing problem is discussed in Section 7 and the scheduling problem is discussed in Section 8. In Section 9 the problem of identifying a critical set of tasks from a solution is addressed. The description of the prototype of DSS we developed is given in Section 10, while finally some computational results can be found in Section 11.

3. CHARACTERISTICS OF THE PROBLEM

In this section we briefly describe and formalize the problem in order to make the paper self contained. A set of projects $\mathbf{P}$ and a set of resources $\mathbf{R}$ are given, where a project P consists of a set of tasks $T(P)$. Let us indicate by $\mathbf{T} := \cup_{P \in \mathbf{P}} T(P)$ the set of all tasks. Tasks require resources for their execution. No preemption of tasks is allowed.

Each resource $R \in \mathbf{R}$ is a renewable machine-type resource and can execute at most one task at a time. For each resource $R \in \mathbf{R}$ several periods of availability can be defined. This means that between two availability periods the resource R is unavailable and therefore no task which requires this resource can be scheduled during the unavailability period.

For each task T there exist $n(T)$ alternative ways of processing the task, called *modes*. Each mode is characterized by a resource set $F \subset \mathbf{R}$ and by a processing time $p(T, F)$. The class of resource sets corresponding to the modes associated with the task T is denoted by $\mathbf{F}(T)$ and $\mathbf{F}(\mathbf{U})$ denotes the set of modes associated with the tasks of U, i.e. $\mathbf{F}(\mathbf{U}) := \cup_{T \in \mathbf{U}} \mathbf{F}(T)$.

In other words the processing of task T requires $p(T, F)$ time units if the resource set F is used and all resources in F are at disposal of the task during its processing. It is expected in a typical situation that the bigger F is the faster the processing is, so what is gained in terms of processing time is lost in terms of resource availability for the other tasks.

In usual project scheduling problems, precedence constraints can be defined between pairs of tasks of the same project. Here we assume that the time spent between the completion of each task T and the beginning of task T' must be included between a minimum and a maximum waiting time $\alpha(T, T')$ and $\beta(T, T')$. This is a generalization of classical precedence constraints. The classical case corresponds to having $\alpha(T, T') = 0$ and $\beta(T, T') = +\infty$, while if $\alpha(T, T') = -\infty$ and $\beta(T, T') = +\infty$ there is no constraint between T and T'. We will say that $T \prec T'$ whenever α and β define a real constraint between T and T'.

A release date $a(T)$ and a deadline $b(T)$ are associated with each task T; these quantities identify the time interval in which the task must be processed. Without loss of generality we assume that $\min_T a(T) = 0$. A due date $d(T)$ can be associated to each task

T, corresponding to the most desired completion time, together with an earliness weight $\nu(T)$ and a tardiness weight $\omega(T)$. The weights ν_{max}, ω_{max}, ν_{sum} and ω_{sum} can be given for the maximum earliness and tardiness and for the total earliness and tardiness.

For each task T we must select a mode $A(T) \in \mathbf{F}(T)$ and a starting time $S(T) \in \mathcal{R}$, where $\mathcal{R}$ is the set of reals. The completion time $C(T)$ of task T is defined as $C(T) := S(T) + p(T, A(T))$. The earliness $V(T)$ and the tardiness $W(T)$ of task T are defined as usual by $V(T) := \max\{0; d(T) - C(T)\}$ and $W(T) := \max\{0; C(T) - d(T)\}$. Maximum weighted earliness V_{max}, maximum weighted tardiness W_{max}, total weighted earliness V_{sum} and total weighted tardiness W_{sum} are defined as

$$V_{max} := \max_{T \in \mathbf{T}} \nu(T)V(T), \qquad W_{max} := \max_{T \in \mathbf{T}} \omega(T)W(T),$$

$$V_{sum} := \sum_{T \in \mathbf{T}} \nu(T)V(T), \qquad W_{sum} := \sum_{T \in \mathbf{T}} \omega(T)W(T).$$

Due to the presence of release dates and deadlines we may model each unavailability interval (t^1, t^2) for the resource R as a dummy task T with $a(T) = t^1$, $d(T) = b(T) = t^2$, with unique associated mode $\{R\}$ and processing time $p(T, \{R\}) = t^2 - t^1$. From now on it will be tacitly understood that the set of tasks also includes the dummy tasks representing unavailability intervals. Therefore we shall never need mentioning explicitly the presence of unavailability intervals.

Given $A(T)$ and $S(T)$ we may consider the following subset

$$H(T) := \big[S(T), \ S(T) + p(T, A(T))\big) \times A(T) \subset \mathcal{R} \times \mathbf{R}$$

associated to task T. Just note that it corresponds to the usual Gantt diagram with time on the horizontal axis and the set of resources on the vertical axis.

The constraints to be imposed on the mode and starting time selection are:

$$
\begin{array}{llll}
H(T) \cap H(T') = \emptyset & \forall T \neq T' & \text{resource constraints} & \\
a(T) \leq S(T) \leq b(T) - p(T, A(T)) & \forall T & \text{time constraints} & (1) \\
\alpha(T, T') \leq S(T') - C(T) \leq \beta(T, T') & \forall T \prec T' & \text{precedence constraints} &
\end{array}
$$

The goal of the scheduling problem is to select a mode and a starting time for each task T so that the objective function

$$Z = \nu_{max}V_{max} + \omega_{max}W_{max} + \nu_{sum}V_{sum} + \omega_{sum}W_{sum} \qquad (2)$$

is minimized, and the constraints (1) are satisfied.

When $n(T) = 1$ and $|F| = 1$, $\forall F \in \mathbf{F}(\mathbf{T})$, and all tasks in a project are linearly ordered, then the problem reduces to a job shop problem. As a generalization of the job shop problem, the problem previously defined, which will be referred to in the following as $\mathcal{P}$, is computationally very hard.

We assume without loss of generality that all data of the problem have integer values.

4. THE DEVELOPMENT OF THE DSS

Our approach to the problem can be described by means of three steps. The first step was to understand the RCSP, to model it in order to provide a sound basis for the design of efficient algorithms. Our attention was focused on the idea of taking advantage of interactivity which we felt as the most interesting feature of a DSS. As our approach to the RCPS turned out to be a decomposition approach and therefore several subproblems were identified, the second step was to work on the subproblems, developing algorithms and starting the implementation phase with the implementation of the different algorithms. Finally, the third step was devoted to the design and implementation of the DSS as a whole, embedding the already implemented parts. Obviously, the three steps were not completely sequential and some feedback was needed too. While we consider the first and second step completed, the implementation of the DSS is still in progress.

Our development process was motivated by our belief that the solution procedures were the heart of our DSS and that the main challenge in the development of the DSS was the design of new interactive and powerful ways to solve a hard, but well defined problem.

We decided to use the Turbo Pascal as an implementation language, because we found it to be a good tool for implementing numerical algorithms and for some graphics. A number of students have been involved in the problem, especially in the software development.

Following our steps in the DSS development, we will first describe our modeling and algorithmic approach and then the DSS.

5. A DECOMPOSITION APPROACH

We decided the problem was too complex to be solved effectively as a whole by any single algorithm. Therefore we have adopted a decomposition approach, both to design good off-line solution procedures and to allow the user to interact at different levels of the solution process. For this second reason we have looked for a decomposition which reflects the natural decision process. It is useful for discussing our decomposition approach to define a schedule in the following way. Let us denote by

$$\mathbf{S} = \bigcup_{I \in 2^{\mathbf{T}}} \mathbf{S}(I)$$

the set of all sequences of all subsets of $\mathbf{T}$, being $\mathbf{S}(I)$ the set of all sequences of tasks in I. A schedule is defined by means of three functions (A, E, S) where

$$A : \mathbf{T} \to \mathbf{F}(\mathbf{T})$$

assigns to each task $T \in \mathbf{T}$ a mode, that is a resource set $F \in \mathbf{F}(T)$ (let $T(R) := \{T \in \mathbf{T} : R \in A(T)\}$),

$$E : \mathbf{R} \to \mathbf{S}$$

assigns to each resource $R \in \mathbf{R}$ a sequence $S \in \mathbf{S}(T(R))$ and

$$S : \mathbf{T} \to \mathcal{R}$$

assigns a starting time $S(T)$. The scheduling problem can be defined as

$$\min_{(A,E,S)} Z(A, E, S) = \nu_{max} V_{max} + \omega_{max} W_{max} + \nu_{sum} V_{sum} + \omega_{sum} W_{sum} \qquad (3)$$

so that

$$
\begin{array}{llll}
A(T) \in \mathbf{F}(T) & \forall T & \text{(assignment constraints)} \\
E(R) \in \mathbf{S}(T(R)) & \forall R & \text{(resource constraints)} \\
a(T) \le S(T) \le b(T) - p(T, A(T)) & \forall T & \text{(time constraints)} \\
\alpha(T, T') \le S(T') - C(T) \le \beta(T, T') & \forall T \prec T' & \text{(precedence constraints)} .
\end{array}
$$

Due to the particular definition of A, E and S, (3) can be rewritten as

$$\min_A \min_E \min_S Z(A, E, S). \qquad (4)$$

Indeed a schedule can be given by first specifying the resource assignment, then specifying the order according to which each resource processes the tasks on the basis of the assigned resource sets and the relative processing times and finally the time assignment on the basis of the resources, the relative processing times and the processing sequences.

We call the problem $\mathcal{P}(AE)$, defined as

$$Z(AE) = \min_S Z(AES),$$

the scheduling problem, consisting in assigning a starting time to each task given a resource set for each task and a processing sequence for each task. The problem defined as

$$Z(A) = \min_E Z(AE)$$

is denoted as $\mathcal{P}(A)$ and will be called the sequencing problem, emphasizing the need of sequencing the tasks on each resource after a mode has been assigned to each resource. Finally, the problem $\mathcal{P}$ can be succintly written as

$$Z = \min_A Z(A). \qquad (5)$$

The formulation (5) suggests that a solution of problem $\mathcal{P}$ can be found by enumerating the resource functions A, then enumerating for each function A the processing functions E and by solving the corresponding problems $\mathcal{P}(A, E)$. Obviously, the computational complexity of the problem makes in general the exhaustive enumeration impossible.

So our approach calls for solving $\mathcal{P}(A, E)$ only for a small subset of possible resource assignments and processing sequences. This subset is found through an iterative process

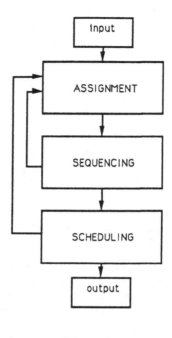

Figure 1

consisting of three phases. In the first phase a tentative assignment A is specified as an optimal assignment for a subproblem which is an approximation for Problem $\mathcal{P}$. This problem is denoted by Problem $\mathcal{A}$. Then a tentative set of processing sequences E is provided as the solution of a subproblem which is an approximation for Problem $\mathcal{P}(A)$. This problem is denoted by Problem $\mathcal{E}$. Finally a time assignment is found by solving optimally Problem $\mathcal{P}(A, E)$, which is also shortly denoted by Problem S.

The main idea of the decomposition is to run the different phases possibly several times by exploiting the information provided by the solutions of Problems $\mathcal{E}$ and S to revise the assignment A (see Figure 1). This can be done by identifying subsets of critical tasks for which changing the mode assignment can improve the objective function.

The problem of how to recover and use information for the identification of a critical set of tasks from a set of processing sequences or from a schedule will be discussed in Section 9.

The procedure described in Figure 1 can be seen as interactive or off-line in dependence of the fact that some or none of the steps of the procedure require interaction with the user. In case of an off-line procedure, a stopping rule for the iterative process must be defined. We will not discuss this issue.

6. THE RESOURCE ASSIGNMENT PROBLEM

In this section we discuss the problem of how to find a mode assignment for a subset of tasks $\mathbf{T'} \subseteq \mathbf{T}$. In other words we discuss the problem of identifying a restricted assignment function $A(T)$ for all $T \in \mathbf{T'}$. A preliminary investigation of this problem can be found in Serafini and Speranza (1988).

The problem of finding such an assignment will be denoted by $\mathcal{A}(\mathbf{T'})$. The case $\mathbf{T'} = \mathbf{T}$ corresponds to the first iteration of the procedure of Figure 1, while the case $\mathbf{T'} \subset \mathbf{T}$ corresponds to any subsequent iteration, when a critical set of tasks $\mathbf{T'}$ has been identified. Let us first consider the case $\mathbf{T'} = \mathbf{T}$. In the case $\mathbf{T'} \neq \mathbf{T}$ the symbol $\mathbf{T}$ should be replaced by $\mathbf{T'}$ in all the present section.

The approach to the assignment problem we present is simply based on the observation that the minimum makespan t_M is greater than the busy time of each resource, that is:

$$t_M \geq \max_{R \in \mathbf{R}} p_R \qquad (6)$$

where p_R denotes the total amount of processing time during which the resource R is actually used. Although the objective can be different from the minimization of the makespan, all objective functions contain a weighted sum of maximum and total tardiness (otherwise all tasks would be scheduled at infinity), which has a strong correlation with the makespan. Moreover, it seems to be a heuristically general principle that the fairest the distribution of the resources among the tasks is the better the schedule is. In production scheduling the machine with the maximum workload is considered as the bottleneck machine and the balancing of the machine workloads is seen as a mean to reach more effective schedules.

In particular, in the case in which there are no precedence constraints among the tasks and each task uses one resource:

$$t_M = \max_{R \in \mathbf{R}} p_R. \qquad (7)$$

Whenever (7) holds, the problem of minimizing the makespan is equivalent to the problem of minimizing $\max_{R \in \mathbf{R}} p_R$. Of course, in the general case there are precedence constraints among the tasks, each task uses more than one resource and the objective can be different from the minimization of the makespan. Therefore the following model, in which the maximum resource use is minimized, only provides a heuristic assignment to problem $\mathcal{P}$.

We may define the following

Problem $\mathcal{A}(\mathbf{T})$:

$$\min_{A} \max_{R \in \mathbf{R}} \sum_{T \in T(R)} p(T, A(T)).$$

Problem $\mathcal{A}(\mathbf{T})$ can be also stated as a 0-1 linear programming problem. Let us define a $|\mathbf{R}| \times \sum_{T \in \mathbf{T}} |\mathbf{F}(T)|$ matrix B, in the following way:

$$b_{RF} := \begin{cases} p(T, F) & \text{if } R \in F, \text{ with } F \in \mathbf{F}(T) \\ 0 & \text{otherwise} \end{cases}$$

and the assignment variables x_F as follows:

$$x_F := \begin{cases} 1 & \text{if } A(T) = F \\ 0 & \text{otherwise} \end{cases}.$$

Thus the assignment model can be stated as follows:

$$y_M = \min y$$
$$\sum_T \sum_{F \in \mathbf{F}(T)} b_{RF} x_F \leq y \qquad \forall R$$
$$\sum_{F \in \mathbf{F}(T)} x_F = 1 \qquad \forall T$$
$$x_F \in \{0,1\} \quad \forall F \in \mathbf{F}(T), T \in \mathbf{T}.$$

On the basis of the above observations, in some very particular cases the solution of problem $\mathcal{A}(\mathbf{T})$ can be used as a test of optimality for a schedule. Indeed, if the objective of the problem is the minimization of the makespan, the resources are always available and it turns out that the makespan t_M of a schedule happens to be equal to y_M, i.e. equation (7) holds, then the schedule is certainly optimal. Obviously, this should be regarded as an exceptional case. In the general case, the resource assignment obtained as a solution of Problem $\mathcal{A}(\mathbf{T})$ does not correspond to the resource assignment of any optimal schedule. In this case, as a result, there must be time intervals in which the resource distribution is unsatisfactory.

Let us observe that different resource assignments can be obtained for the same task set $\mathbf{T}$ changing the objective function. A possible general form of the objective function of problem $\mathcal{A}(\mathbf{T})$ could be

$$\min k_1 f_1(y) + k_2 f_2(x),$$

where $f_1(y)$ and $f_2(x)$ are generic functions and k_1 and k_2 are weighting parameters. At the present moment we experienced and tested only the first form of the objective function and we will not discuss this general form in the following. However, we consider this as one of the possibilities to investigate in order to improve our DSS.

For instance if we consider the following alternative minimization:

$$\min_A \sum_{R \in \mathbf{R}} \sum_{T(R)} p(T, A(T))$$

the resulting problem can be trivially solved by finding for each task T the mode which minimizes

$$\{|F| \cdot p(T, A(T)) : F \in \mathbf{F}(T)\}.$$

This objective function has the practical justification of measuring the total time resources are busy. However it is not sensitive to the interaction of tasks for resource competition. Hence it provides the same solution independently of the task set $\mathbf{T}$. Since it has the undoubtful advantage of being easy to compute we shall use this criterion together with other heuristic procedures.

In order to understand the resource assignment problem let us first note that the problem $\mathcal{A}(\mathbf{T})$ tries to settle the competition of tasks for the same resources all over the time horizon of the process. Actually this is in some case unnecessary because tasks whose starting times are far apart in time do not really compete even if they share some resources. Intuitively two tasks do compete for a common resource only if they are so close in time that their processing times tend to overlap, but cannot do it because of the common resource and so they are forced to a disjunctive precedence constraint. Furthermore due to the presence of release dates and deadlines, certain classes of tasks are always separated in time.

These considerations lead us to focus our attention to specially selected time intervals, called *time windows* in the sequel, and therefore to solve again an assignment problem, which is in this case restricted to a subset of tasks whose processing times, as given by the sequencing or the scheduling problems, overlap with the time window. The problem of identifying such time windows will be discussed in Section 9.

Furthermore not all tasks within a time window have the same effect on the objective function value. For instance if the objective function takes into account only the maximum tardiness, there are tasks on the critical path for which a change in the processing time has an immediate effect on the maximum tardiness, whereas other non critical tasks may be delayed without affecting the maximum tardiness. Hence we should attach more importance to tasks which are on the critical path. We postpone to Section 9 a discussion of how to evaluate the relative importance of tasks as it can be inferred from the solutions of Problems $\mathcal{E}$ and $\mathcal{S}$. Here we only say that each task T is assigned a weight $0 \leq \gamma(T) \leq 1$. This weight is used to rescale the processing times according to

$$p'(T, A(T)) := \gamma(T) \cdot p(T, A(T)) \qquad \forall A(T) \quad \forall T.$$

Problem $\mathcal{A}(T)$ is then solved for all tasks in $\mathbf{T}$ by using the processing times p'.

Now we characterize the assignment problem $\mathcal{A}(\mathbf{T})$ in order to design algorithms for its solution. First note that the particular case in which each task T can be processed by any (singleton) resource with the same processing time for all resources (so that $|F| = 1, \forall F \in \mathbf{F}(T), \cup_{F \in \mathbf{F}(T)} F = \mathbf{R}$ and $p(T, F) = p_T, \forall F \in \mathbf{F}(T))$, the resources are always available and the objective is the minimization of the makespan, is the *Bin Packing Problem*, with each resource interpreted as a 'bin' and the processing time the size of the 'item' T to be inserted into some bin (see [8]).

Therefore Problem $\mathcal{A}(T)$ is NP-hard and there is no hope for an efficient algorithm. Let us consider the following relaxed Problem $\overline{\mathcal{A}(\mathbf{T})}$

$$\overline{y}_M \;\; = \;\; \min y$$

$$\sum_{T} \sum_{F \in \mathbf{F}(T)} b_{RF} x_F \leq y \qquad \forall R$$

$$\sum_{F \in \mathbf{F}(T)} x_F = 1 \qquad \forall T$$

$$x_F \geq 0 \qquad F \in \mathbf{F}(T), T \in \mathbf{T}.$$

The optimal solution of $\overline{\mathcal{A}(\mathbf{T})}$ is not integral in general, so its assignment variables x_F do not define an assignment for those tasks T for which there exists $F \in \mathbf{F}(T)$ such that

$0 < x_F < 1$, i.e. x_F is fractional. In the following we investigate the number of integral solutions of $\mathcal{A}(\mathbf{T})$.

First note that the above linear programming problem, once it has been converted into standard form, has $\sum_T |\mathbf{F}(T)| + |\mathbf{R}| + 1$ variables (including slacks) and $|\mathbf{R}| + |\mathbf{T}|$ rows. Let k be the number of active constraints in the optimal solution of $\overline{\mathcal{A}(\mathbf{T})}$. Note that the variable y, being unconstrained, must be in any basis, so that $|\mathbf{R}| + |\mathbf{T}| - 1$ variables, among the assignment and the slack ones, must be in any basis. The slack variables corresponding to non active constraints must be in the optimal basis, so at most

$$|\mathbf{R}| + |\mathbf{T}| - 1 - (|\mathbf{R}| - k) = |\mathbf{T}| + k - 1$$

assignment variables are in the optimal basis.

Note also that the assignment constraints imply that at least one assignment variable per task must be strictly positive and thus in the optimal basis. It follows that for at most $(k-1)$ tasks there are fractional assignment variables and consequently the optimal solution of $\overline{\mathcal{A}(\mathbf{T})}$ is an assignment for at least $(|\mathbf{T}| - k + 1)$ tasks. Hence at most

$$|\mathbf{T}| + k - 1 - (|\mathbf{T}| - k + 1) = 2k - 2$$

optimal assignment solutions of $\overline{\mathcal{A}(\mathbf{T})}$ are fractional.

It is interesting to consider the dual problem of $\overline{\mathcal{A}(\mathbf{T})}$

$$\overline{y}_M = \max \sum_{T \in \mathbf{T}} v(T)$$

$$v(T) \leq p(T, F) \cdot \sum_{R \in F} u(R) \qquad \forall F \in \mathbf{F}(T), \forall T \in \mathbf{T}$$

$$\sum_{R \in \mathbf{R}} u(R) = 1$$

$$u(R) \geq 0$$

Denoting $\bar{u}(F) := \sum_{R \in F} \bar{u}(R)$, with $\bar{u}(R)$ the optimal dual relative to the resource constraints, the optimal linear programming value can be written as

$$\overline{y}_M = \sum_{T \in \mathbf{T}} \min_{F \in \mathbf{F}(T)} p(T, F) \bar{u}(F) \tag{8}$$

Equation (8) has a direct interpretation: each variable $\bar{u}(R)$ measures the scarcity of the resource R with respect to the other resources. Note that $0 \leq \bar{u}(R) \leq 1$, so this measure is expressed as percentage. A value $\bar{u}(R) = 0$ means that the use of resource R does not affect the optimal value and so the scarcity of R is null. On the contrary $\bar{u}(R) = 1$ means that the resource R is fully responsible for the optimal value (note that in this case all other optimal duals $\bar{u}$ are equal to zero because of the constraint $\sum \bar{u}(R) = 1$).

By summing $\bar{u}(R)$ over the resources actually employed by the mode which uses the resource set F for task T, we get the quantity $\bar{u}(F)$ which may be called the *utilization factor* of the resource set F. It is as if all resources had been subsumed by one single fictitious resource and the mode corresponding with the resource set F used this resource at a level given by the utilization factor $\bar{u}(F)$. The processing time of the fictitious resource applied to the resource set F for task T is given by the product of the 'true' processing time $p(T, F)$ times the utilization factor $\bar{u}(F)$.

Then, among the resource sets $F \in \mathbf{F}(T)$ one has to choose the one minimizing the fictitious processing time $\bar{u}(F)p(F)$. If this minimum is unique then the corresponding primal variables are integral and give rise to an assignment.

We designed and included in the DSS several solution procedures for the resource assignment problem, which are inspired by different algorithmic paradigms. Here we give a short description of them.

The complexity of the resource assignment problem together with the fact that the problem is only a subproblem of an iterative solution approach makes it particularly interesting the design of heuristic algorithms. However, in the solution procedures we included a branch-and-bound algorithm, which can be used for the solution of small sized problems. Moreover, a truncation of the resulting search tree makes it a basis of a good heuristic procedure, as it will be shown.

Let us briefly describe the branch-and-bound algorithm. The search tree has levels corresponding to the tasks and the branches outgoing from a node correspond to the modes of a certain task. Therefore each node of the search tree is characterized by the fact that some tasks have forced assignments. The successors of a node inherit the same forced asssignments and a task, not yet assigned, receives forced assignments to different modes for all successors. The root of the search tree has no forced assignments.

Denoting by $\mathbf{T_0}$ the set of forcedly assigned tasks and by $\mathbf{T_1}$ the other tasks in a certain node s of the search tree, and by $\mathbf{F}_s(\mathbf{T_0})$ the forced assignments corresponding to node s, the following linear programming problem has to be solved on s:

$$\bar{y}_M(s) \;=\; \min \; y$$

$$\sum_{F \in \mathbf{F}(\mathbf{T_1})} b_{RF} x_F \leq y - \sum_{F \in \mathbf{F}_s(\mathbf{T_0})} b_{RF} \quad \forall R$$

$$\sum_{F \in \mathbf{F}(T)} x_F = 1 \qquad\qquad \forall T \in \mathbf{T_1}$$

$$x_F \geq 0 \qquad\qquad \forall F \in \mathbf{F}(\mathbf{T_1})$$

The first computational experience shows that it is convenient to branch over that task whose solution is most fractional, i.e. such that $\max_{F \in \mathbf{F}(T)} x_F$ is minimal.

The heuristic truncation of the search tree is obtained by keeping fixed the values of those variables that are integral in the previously solved linear programming problems. According to the considerations of the previous section we are left with a smaller system of at most $2|\mathbf{R}| - 2$ assignment variables and $|\mathbf{R}| - 1$ tasks. Although this does not guarantee optimality of the final solution, it definitely speeds up the computation.

Several greedy algorithms have been included in the DSS which are also used to identify a starting solution for other procedures, which will be described in the following. The greedy procedures differ for the criterion according to which a mode is assigned to each task among the available ones and some of them are inspired to the well known greedy procedures for the bin packing problem. The greedy algorithms process the tasks in non decreasing order of number of modes $n(T)$. In particular:

MinSum: The mode is selected which minimizes the product of the processing time times the number of resources.

FirstSmaller: The mode corresponding to the shortest processing time (in case of ties, the first mode) is selected.

FirstRandom: The mode is selected randomly.

FirstBigger: The mode corresponding to the longest processing time (in case of ties, the first mode) is selected.

LowestFit: The mode which minimizes the maximum partial resource usage is selected. In other words, denoting by $\overline{y}_R$ the usage of resource R due to the already examined tasks, the mode for task T is selected such that

$$\min_{F \in \mathbb{F}(T)} \max_{R \in F} p(T, F) + \overline{y}_R.$$

In case of ties, the modes may be selected in different ways. In particular we have:

LowestFitMin: In case of ties the mode with the shortest processing time is selected.

LowestFitMax: In case of ties the mode with the longest processing time is selected.

LowestFitLev: In case of ties the mode which minimizes the number of times a resource is used is selected.

The solutions obtained by the greedy procedures can be accepted as satisfactory or can be possibly improved by means of local searches. The following local search algorithm has been included in the DSS:

k-change: The algorithm is parametric in k, whose value must be provided by the user. Among all possible solutions obtained by changing k modes the best one is selected.

Finally, a simulated annealing procedure has been also designed in which a neighbour of a solution is defined as a solution which differs for a single mode.

As a further approach to the assignment problem we would like to mention an alternative method which takes care directly of time constraints present in the problem data. However, this approach is not amenable to mathematical programming techniques and special algorithms should be designed for its solution. This will be matter of future reasearch. At the moment we only formulate the problem.

The idea is the following. For each assignment we build a diagram for each resource R measuring the amount of demanded resource units as a function of time. This amount is obtained by positioning the completion time of each task requiring the resource R exactly at its due date. So we have starting times $d(T) - p(T, A(T))$ and completion times $d(T)$ and we may build the following function

$$f_R(A, t) := \# \text{ of tasks in } T(R) \text{ started in } (-\infty, t) - \# \text{ of tasks in } T(R) \text{ completed in } (-\infty, t].$$

If we conventionally put equal to zero the current time we may note that for positive t one resource unit only is available for each resource and so we have to weight the excess of this function with respect to one, i.e. we should consider the following objective function $\max\{f_R(A,t) - 1; 0\}$. However, for negative t no resource units are available at all since no task can be scheduled for processing in the past and therefore we have to consider the objective function $f_R(A,t)$. Taking into account all these observations the attempt should be directed toward minimizing over all possible assignments A

$$\max_R \int_{-\infty}^0 f_R(A,t)\,dt + \int_0^{+\infty} \max\{f_R(A,t) - 1; 0\}\,dt.$$

Let us remark that in case $d(T) = 0$ for all possible tasks we have exactly the formulation which has been discussed in this section.

7. THE SEQUENCING PROBLEM

In this section we consider how to sequence the tasks for each resource once the assignment of tasks to modes has been carried out. The problem to be solved looks like an extension of the well known Job Shop Problem. A preliminary investigation of this type of problems can be found in Serafini and Ukovich (1988) and Serafini et al. (1989). In this paper only an overview is provided of the basic issues of the problem and the relative algorithms. A more detailed analysis of the algorithms described in this section can be found in Lancia (1990).

Formally the problem we address can be formulated as a graph problem whose data are derived from the data of Problem $\mathcal{P}$ and the solution of Problem $\mathcal{A}$. So we are given

- a finite set $\mathbf{T}$ of vertices (tasks);
- a set C of directed arcs, called *conjunctive* arcs, taking care of the precedences $T \prec T'$, such that the directed graph $(\mathbf{T}, C)$ is acyclic;
- subsets $A(T) \subset \mathbf{R}$, for each T, corresponding to the mode assignments;
- nonnegative real values $p(T)$, $\forall T$, corresponding to the processing times of the tasks. For ease of notation we write $p(T)$ instead of $p(T, A(T))$, since the modes are fixed for this problem;
- real values $a(T)$, $\forall T$, corresponding to the release dates;
- real values $b(T)$, $\forall T$, corresponding to the deadlines;
- real values $d(T)$, $\forall T$, corresponding to the due dates.

The variable data to be determined are:

- a set $S = \cup_{R \in \mathbf{R}} S_R$ of linear orderings S_R, called *complete selection*, for the subsets $T(R) \subset \mathbf{T}$, i.e. $S_R \in \mathbf{S}(T(R))$, labeling the tasks of each subset $T(R)$ as $T_{R,1}, \ldots, T_{R,|T(R)|}$. Correspondingly the following set $D(S)$ of directed arcs, called *disjunctive* arcs, is defined

$$D(S) := \bigcup \{(T_{R,i}, T_{R,i+1}) : R \in \mathbf{R}; \ i := 1, \ldots, |T(R)| - 1\}.$$

Let $E(S) := C \cup D(S)$. The graph $G(S) := (\mathbf{T}, E(S))$ is called the *disjunctive graph* relative to the selection S. A selection is called *partial* if the linear orderings are specified only for a proper subset of resources. A selection is said to be *feasible* if the graph $G(S)$ is acyclic. Note that parallel directed arcs may exist.

Given a feasible complete selection we may compute a schedule $S(T)$ by dynamic programming techniques. The schedule $S(T)$ is said to be feasible if

$$S(T) \geq a(T) \qquad \forall T \in \mathbf{T} \tag{9}$$
$$S(T'') - S(T') \geq p(T') \qquad \forall (T', T'') \in E(\mathcal{S}) \tag{10}$$
$$S(T) + p(T) \leq b(T) \qquad \forall T \in \mathbf{T}. \tag{11}$$

The dynamic programming technique may compute $S(T)$ as either earliest or latest starting times for the tasks. In the former case the schedule is automatically feasible with respect to (9) and (10), whilst in the latter case it is automatically feasible with respect to (10) and (11). Both computations give also the critical paths which are the longest paths $\Pi(T, T') : T \to T'$ with length $L(T, T')$ exceeding the value $b(T') - a(T)$ (i.e. in case of infeasibility). We may also define a 'most' critical path $\hat{\Pi}$ as the one for which the quantity $b(T') - a(T) - L(T, T')$ is maximum. Other criteria could be used as well to identify a 'most' critical path.

The objective of Problem $\mathcal{E}$ is finding a complete selection and a correspondingly feasible schedule such that $\nu_{max} V_{max} + \omega_{max} W_{max}$ is minimized, as an approximation for Problem $\mathcal{P}(A)$.

Precedence constraints (see (1)) can be dealt with in Problem $\mathcal{E}$ if $\alpha(T, T') \geq 0$ and $\beta(T, T') = +\infty$. The case $\alpha(T, T') = 0$ (and $\beta(T, T') = +\infty$) corresponds to a usual precedence and is taken care of by Problem $\mathcal{E}$ via the directed arcs C. The case $\alpha(T, T') > 0$ can be also taken care of by Problem $\mathcal{E}$ by adding a dummy task $\hat{T}$ (with no resource involved) of processing time equal to $\alpha(T, T')$ with usual precedences $T \prec \hat{T} \prec T'$. The case with generic values for $\alpha(T, T')$ and $\beta(T, T')$ cannot be modelled adequately in Problem $\mathcal{E}$ and so we consider this type of constraints in Problem $\mathcal{S}$.

Our approach relies on solving several times simpler problems involving one resource only. The first of these problems is defined as follows (Two Sided One Machine Problem, TSOMP):

A set $T(R)$ of tasks is given. For each $T \in T(R)$ the integer quantities $aa(T)$, $p(T)$, $bb(T)$, $q^-(T)$ and $q^+(T)$ are assigned. The values $q^+(T)$ and $q^-(T)$ are called *tails* and *heads*. A partial order $\prec_*$ among the tasks is given. The problem consists in finding a permutation $T_1, \ldots, T_{|T(R)|}$ of the tasks and a schedule $S : T(R) \to Z$ such that

$$S(T) \geq aa(T) \qquad \forall T \in T(R) \tag{12}$$
$$S(T) + p(T) \leq bb(T) \qquad \forall T \in T(R) \tag{13}$$
$$S(T') \geq S(T) + p(T) \qquad \forall T \prec_* T' \tag{14}$$
$$S(T_{i+1}) \geq S(T_i) + p(T_i) \qquad i := 1, \ldots, |T(R)| - 1 \tag{15}$$

and the objective function

$$\max_{T, T' \in T(R)} \left\{ S(T) + p(T) + q^+(T) - S(T') + q^-(T') \right\} \tag{16}$$

is minimized. Let M be the optimal value.

We briefly comment on the TSOMP. First note that the objective function (16) is invariant with respect to a uniform shift of the schedule $S(T)$. Therefore, given an optimal schedule, there may exist other optimal schedules defined by a uniform shift as long as these are feasible with respect to (12) and (13). Second (16) is equivalent to minimizing

$$\max_{T \in T(R)} \left\{ S(T) + p(T) + q^+(T) \right\} + \max_{T \in T(R)} \left\{ q^-(T) - S(T) \right\}$$

i.e.

$$\max_{T \in T(R)} \left\{ S(T) + p(T) + q^+(T) - B \right\} + \max_{T \in T(R)} \left\{ q^-(T) - S(T) \right\} \qquad (17)$$

where B is arbitrary. Minimizing (17) is also equivalent to minimizing

$$\max_{T \in T(R)} \left\{ S(T) + p(T) + q^+(T) - B \; ; \; 0 \right\} + \max_{T \in T(R)} \left\{ q^-(T) - S(T) \; ; \; 0 \right\} \qquad (18)$$

if there is a shift such that both terms are nonnegative. In other words there is no charge for a task T to be scheduled between $q^-(T)$ and $B - q^+(T) - p(T)$ in the case $q^-(T) \leq B - q^+(T) - p(T)$ and there is a constant charge equal to $q^-(T) + p(T) + q^+(T) - B$ for a task T to be scheduled between $B - q^+(T) - p(T)$ and $q^-(T)$ in the case $q^-(T) \geq B - q^+(T) - p(T)$. If the task has to be scheduled outside these intervals there is an increasing penalty and the objective function takes care of the maximum penalty. The idea is to set the values for $q^-(T)$ and $q^+(T)$ so as to force the tasks to be scheduled as much as possible in correspondence to their due dates taking also into account the relative importance of the tasks.

The data for the TSOMP are derived from the data of Problem $\mathcal{E}$ as follows: $T \prec_* T'$ whenever there exists a directed path in $G(S)$ from T to T', with S a partial selection. Let $L(T, T')$ be the length of the longest directed path in $G(S)$ computed by excluding both $p(T)$ and $p(T')$. Put conventionally $L(T, T') := 0$ if there is no such a path and $L(T, T) := 0$. Then $aa(T)$ and $bb(T)$ are computed as

$$aa(T) := \max \left\{ a(T) \; ; \; \max_{T' : T' \prec_* T} \left\{ a(T') + p(T') + L(T', T) \right\} \right\}$$

$$bb(T) := \min \left\{ b(T) \; ; \; \min_{T' : T \prec_* T'} \left\{ b(T') - p(T') - L(T, T') \right\} \right\}.$$

Let $B := \max_T bb(T)$. The quantities $aa(T)$ and $bb(T)$ are called *effective* release dates and deadlines because they embed the effect on T of release dates and deadlines of other tasks in $T(R)$ due to the precedence constraints present in the graph $G(S)$. This guarantees that a feasible schedule with respect to the TSOMP turns into a feasible schedule for Problem $\mathcal{P}$ once the sequence provided by the TSOMP is fed back into $G(S)$.

Now we consider how earliness and tardiness minimization for each task affect the scheduling of tasks in $T(R)$. To this aim we introduce the concept of effective due dates. We may define *earliness-effective* due dates $d^-(T)$ and *tardiness-effective* due dates $d^+(T)$ as follows

$$d^-(T) := \max \left\{ d(T) \; ; \; \max_{(T',T)\in E(S)} d^-(T') + p(T) \right\}$$

$$d^+(T) := \min \left\{ d(T) \; ; \; \min_{(T,T')\in E(S)} d^+(T') - p(T') \right\}.$$

Note that by definition $d^+(T) \leq d(T) \leq d^-(T)$. From these two values we define *earliness heads* $q_1^-(T)$ and *tardiness tails* $q_1^+(T)$ as

$$q_1^-(T) := d^-(T) - p(T), \qquad q_1^+(T) := B - d^+(T)$$

Moreover we may define *feasibility heads* $q_0^-(T)$ and *feasibility tails* $q_0^+(T)$ as

$$q_0^-(T) := aa(T), \qquad q_0^+(T) := B - bb(T).$$

Note that by the definitions $q_0^-(T) \leq q_1^-(T)$ and $q_0^+(T) \leq q_1^+(T)$. Finally we derive the head and tail values to be used in the TSOMP by a convex combination of the above defined heads and tails:

$$q^-(T) := q_0^-(T) + \lfloor (q_1^-(T) - q_0^-(T))\sigma^-(T) \rfloor$$
$$q^+(T) := q_0^+(T) + \lfloor (q_1^+(T) - q_0^+(T))\sigma^+(T) \rfloor$$

where the weights $0 \leq \sigma^-(T), \sigma^+(T) \leq 1$ have to be set according to the relative importance of earliness and tardiness minimization.

In order to be more specific let us make the following observations. If we take $q^-(T) := q_0^-(T)$ and $q^+(T) := q_0^+(T)$ we only take into account release dates and deadlines and the objective function (16) tries to find a feasible schedule, as it happens if the optimal value M of the TSOMP is $M \leq B$. Indeed if there is a feasible schedule $S(T)$ then it follows immediately from the various definitions that $M \leq B$. Conversely let us suppose that $M \leq B$. We first recall that the objective function is invariant with respect to a uniform shift of all $S(T)$. Thus since by hypothesis there exists a schedule $S(T)$ such that $S(T) + p(T) - bb(T) - S(T') + aa(T') \leq 0, \forall T, T'$, there always exists a shift such that both $S(T) + p(T) - bb(T) \leq 0$ and $-S(T') + aa(T') \leq 0$, i.e. the schedule is feasible. Note that if there is no feasible schedule we may arbitrarily choose whether to violate release date constraints (12) or deadline constraints (13) (or both). Of course we assume that the constraints (14) and (15) cannot be violated.

Now, raising the values of the heads from $q_0^-(T)$ to $q_1^-(T)$ corresponds to binding more strictly the starting times of the tasks from below and raising the values of the tails from $q_0^+(T)$ to $q_1^+(T)$ does the same from above. For instance if we take $q^-(T) := q_0^-(T)$ and $q^+(T) := q_1^+(T)$ this choice has the effect of pushing back all tasks in order to minimize the tardiness and to have them at the same time feasible with respect to the release dates. On the contrary the choice $q^-(T) := q_1^-(T)$ and $q^+(T) := q_0^+(T)$ has the effect of pulling forth all tasks in order to minimize the earliness and to have them at the same time feasible with respect to the deadlines.

Note that by the choice $q^-(T) := q_1^-(T)$ and $q^+(T) := q_1^+(T)$ the condition expressing equivalence of the objective functions (17) and (18) is satisfied.

In view of the previous observations a possible choice for the weights σ could be the following (let $\upsilon := \max\{\nu_{max}; \omega_{max}\}$, $\bar{\nu} := \max_T \nu(T)$ and $\bar{\omega} := \max_T \omega(T)$):

$$\sigma^-(T) := \frac{\nu(T)}{\bar{\nu}}\frac{\nu_{max}}{\upsilon}, \qquad \sigma^+(T) := \frac{\omega(T)}{\bar{\omega}}\frac{\omega_{max}}{\upsilon}.$$

We approximately solve the TSOMP by solving a number of times a simpler version of this problem, which is called the One Machine Problem (OMP). It can be shown that this approach is exact if either $\nu_{max} = 0$ or $\omega_{max} = 0$ and $\nu(T) = \omega(T) = 1$, $\forall T$. Formally the OMP is defined by a set $T(R)$ of tasks and for each $T \in T(R)$ the integer quantities $r(T)$, $p(T)$, $q(T)$ are assigned. The problem consists in finding a permutation $T_1, \ldots, T_{|T(R)|}$ of the tasks and a schedule $S : T(R) \to Z$ such that

$$\begin{aligned} S(T) \geq r(T) \qquad & T \in T(R) \\ S(T_{i+1}) \geq S(T_i) + p(T_i) \qquad & i := 1, \ldots, |T(R)| - 1 \end{aligned} \qquad (19)$$

and the makespan

$$\max_{T \in T(R)} \{S(T) + p(T) + q(T)\} \qquad (20)$$

is minimized. Even this simpler version of one machine problems is strongly NP-hard (Garey and Johnson, 1979, p. 102). However, its simpler structure allows an exact algorithm (Carlier, 1982) which is very fast on the average. Note that the OMP can be equivalently stated by dropping the constraint (19) and replacing the objective function (20) by

$$\max_{T,T'} \{S(T) + p(T) + q(T) - S(T') + r(T')\}$$

so that it closely resembles the TSOMP.

The data for the OMP are derived by guessing a value for the parameter ζ from which values for $r(T)$ and $q(T)$ are computed in the following way:

$$\begin{aligned} r(T) &:= \max\left\{q_0^-(T)\,;\, q^-(T) - \left\lceil\frac{\zeta}{\nu_{max}\nu(T)}\right\rceil\right\} \\ q(T) &:= \max\left\{q_0^+(T)\,;\, q^+(T) - \left\lceil\frac{\zeta}{\omega_{max}\omega(T)}\right\rceil\right\}. \end{aligned} \qquad (21)$$

The schedule provided by the OMP is automatically feasible with respect to the effective release dates, since $S(T) \geq r(T) \geq q_0^-(T) = aa(T)$. If it is not feasible with respect to the effective deadlines, it may be possible to turn it into a feasible schedule by decreasing the starting time of some tasks up to $S(T) = aa(T)$. According to whether the schedule provided by the OMP is feasible (with possible shift), it is possible to make a new guess for ζ in a binary search fashion. The range of values for the binary search is given by

$$0 \leq \zeta \leq \zeta_M := \max_T \{\nu_{max}\nu(T)\left(q^-(T) - q_0^-(T)\right)\,;\, \omega_{max}\omega(T)\left(q^+(T) - q_0^+(T)\right)\}$$

where the value ζ_M is computed from (21). If for a value ζ the solution is feasible with respect to the TSOMP then the new guess is $\zeta := \zeta/2$, otherwise it is $\zeta := (\zeta + \zeta_M)/2$. If there is no feasible solution for $\zeta = \zeta_M$ (i.e. for $r(T) = q_0^-(T)$ and $q(T) = q_0^+(T)$) then the TSOMP is not feasible. Raising the value of ζ from 0 to ζ_M has the effect of relaxing the goal of optimizing the objective function in order to force feasibility.

Until now there has been a full simmetry with respect to both directions of time. Now we are going to break this simmetry. Indeed we adopt the point of view that release date constraints are more compelling than deadline constraints. This is motivated by physical reasons since in reality there is no simmetry of time and therefore while it is possible to violate a deadline simply by 'waiting', it is typically difficult to violate a release date by anticipating events. We have also to consider that release dates are usually external constraints not under the control of the decision maker of the project and so there is little hope to anticipate events.

Therefore in the sequel we shall adopt the point of view that if a schedule has to be infeasible it is infeasible with respect to the deadline constraints only. Violation of the constraints is measured as

$$\max_T \left\{ \omega(T) \cdot \max \left\{ S(T) + p(T) - bb(T) \, ; \, 0 \right\} \right\}.$$

The TSOMP can be used to find good sequences for the Problem $\mathcal{E}$. Indeed the Shifting Bottleneck Procedure (SBP) is strongly based on this idea. The SBP is fully described in Adams et al. (1988). Here we outline its general scheme which can be described by the following high level algorithm:

```
procedure SBP;
begin
      I := ∅;
      Q := R;
      while Q ≠ ∅ do
      begin
            choose R ∈ Q;                          {choice phase}
            transfer R from Q to I;
            reschedule(I);                         {reoptimization phase}
      end
end.
```

The crucial operations are the choice of a particular resource from $\mathbf{Q}$ and the rescheduling of $\mathbf{I}$. Priority is given to the most critical resource (whence the name 'bottleneck'). Here the measure of criticality for the resource $R \in \mathbf{Q}$ is obtained by solving a TSOMP. The TSOMP may be feasible or not. We rank the resources in a lexicographic order with respect to the two criteria of deadline violation and makespan. Priority is given to the deadline violation. This ranking corresponds to a decreasing criticality ordering of the resources.

The same criterion is employed in order to reschedule I. The high level algorithm for the procedure reschedule(I) is the following:

```
procedure reschedule(I);
begin
        for i := 1 to K do
        begin
                for j := 1 to |I| do
                begin
                        solve a TSOMP for R_j w.r.t. the graph (T, E(S_{I\R_j}));
                        resequence R_j according to the TSOMP solution
                end
                if changesort then sort I in order of decreasing criticality
                        as R_1, R_2, ... R_{|I|};
        end
end.
```

The parameter K has to be set during implementation. In Adams et al. (1988) the value $K = 3$ has been suggested, except for the last loop when the iteration is continued until no improvement can be found. However it has been also found that a value $K = 1$ does not disprove too much the quality of the solution while speeding up the computation. Let us note again that the order in which the resources are selected depends on the previously computed makespan. The resource just entering the set I has as a measure of criticality the value computed for the insertion into I. The boolean parameter $changesort$ has also to be set during implementation. As can be seen from the procedure if it is set to 'false' the resources in I are always rescheduled in the same order in which they entered I. This parameter is not considered in Adams et al. (1988) where it is implicitly set to 'true', thereby justifying the term 'shifting' in the procedure name. However, extensive experimentation has shown that sometimes setting the parameter to 'false' gives better results.

We recall that the sequencing problem is concerned with all tasks only in the first step. In subsequent steps, after the reassignment of modes to some tasks, only a partial set of resources need to be resequenced. This makes the SBP much faster since it is like starting the SBP with some non empty subset I.

Another point concerns the One Machine Problems. It has been remarked that they can be solved quite fast by the Carlier's algorithm. However, they are NP-hard and as such it is not surprising that once in a while some instances appear with very long computing times. It has been observed that these troublesome instances can take up to half of the total computing time for Problem $\mathcal{E}$. In these cases it is better to give up optimality of the One Machine Problem in order not to waste time. We suggest to put a bound on the number of subproblems generated by Carlier's algorithm (we recall it is a branch and bound method). We have found through experimentation that this bound can be set to 50. In general it is convenient to leave it as a parameter to be tuned during implementation.

8. THE SCHEDULING OF THE TASKS

Once the modes have been assigned to the tasks and for each resource the time sequence of its tasks has been defined, we are left with Problem $\mathcal{P}(A, E)$, i.e. Problem S, whose contraints and objective function can be modelled in the framework of minimum cost network flow problems. We first recall some theoretical facts about network flows which will be useful in the sequel.

Two functions may be defined on a network (see Rockafellar, 1984, and Lancia, 1990): a primal *flow* function $x : E \to R$ (with E the arc set) and a dual *potential* function $u : N \to R$ (with N the node set). The derived function $v : E \to R$ defined as $v_e := u_j - u_i$ with $e = (i, j)$ is called *tension*. For each arc e the flow x_e is constrained to lie within a *capacity* interval $[c_e^-, c_e^+]$ and the tension v_e is constrained to lie within a *span* interval $[d_e^-, d_e^+]$ (both intervals may be possibly unbounded). On a network we may define the following

Primal Problem P:

$$\phi := \min \quad \Phi(x) := \sum_{e \in E} f_e(x_e)$$

$$Ax = b$$

where A is the node-arc incidence matrix of the network, b is a given vector (this b has nothing to do with $b(T)$ denoting deadlines - since we shall only use the value $b = 0$ no confusion should arise) and each $f_e(x_e) : R \to R \cup +\infty$ is a convex function of the flow on the arc e only. These functions also embed the capacity constraints on the individual arc flows. A flow x is said to be *feasible* if $Ax = b$ and $\Phi(x) < +\infty$. From Problem P we may derive the dual function $\Psi(u, v)$ through the Lagrangian function $\Phi(x) + u(Ax - b)$ as

$$\Psi(u, v) := \inf_x \Phi(x) + u(Ax - b) = -\sum_{i \in N} u_i b_i + \sum_{e \in E} \inf_{x_e} (f_e(x_e) - v_e x_e) =$$

$$-\sum_{i \in N} u_i b_i - \sum_{e \in E} \sup_{x_e} (v_e x_e - f_e(x_e)) = -\sum_{i \in N} u_i b_i - \sum_{e \in E} g_e(v_e),$$

where $g(v) := \sup_x (vx - f(x))$ is called the *conjugate* function of $f(x)$ and satisfies also the relationship $f(x) = \sup_v (vx - g(v))$ (i.e. f is also the conjugate function of g). Then we may define the following

Dual Problem D:

$$\psi := \sup \quad \Psi(u, v)$$

$$v = -uA.$$

Note that the functions $g_e(v)$ may embed span constraints. A tension v is said to be feasible if there exists u such that $v = -uA$ and $\Psi(u, v) > -\infty$. Moreover the important result $\psi = \phi$, called *strong duality*, holds under very mild assumptions. For instance it can be proven that if Φ is a piecewise linear convex function and there exists a feasible flow then strong duality holds. For each $e \in E$ we may define the following curve $\Gamma_e \subset R^2$ which is called the *characteristic curve* of e

$$\Gamma_e := \{(x_e, v_e) : f_e^-(x_e) \le v_e \le f_e^+(x_e)\} \tag{22}$$

or alternatively

$$\Gamma_e := \{(x_e, v_e) : g_e^-(v_e) \le x_e \le g_e^+(v_e)\}.$$

Just note that each Γ_e curve is simply the graph of the derivative of the function f_e plus all vertical segments drawn to 'fill' the discontinuity jumps. So, since each f_e is convex, each Γ_e curve is monotone in the sense that $(x, v) + R_+^2 + R_-^2 \supset \Gamma, \forall (x, v) \in \Gamma$. In the case of piecewise linear functions f_e, the Γ_e curve is stepwise. A fundamental result of network flow theory states that $(\hat{x}, \hat{v})$ is optimal if and only if $(\hat{x}_e, \hat{v}_e) \in \Gamma_e, \forall e$.

The problem we are dealing with, that is Problem S, can be modelled as a dual network flow problem in which the potentials play the role of starting times of the tasks. The network we build has nodes corresponding to the tasks denoted by T plus three distinguished nodes called source, tardiness and earliness nodes, denoted by T^0, T^+ and T^- respectively. Conventionally and without loss of generality we assign potential $u(T^0) := 0$. The arcs of the network represent constraints between pairs of tasks which are modelled here as piecewise linear convex dual objective functions. As it will be seen in the sequel there is no need of weighting the potentials u_i through the weights b_i, so we assign $b = 0$ and the flow we are looking for is actually a circulation.Reconsidering the constraints (18) we have to deal with resource, time and precedence constraints.

Resource constraints, once the sequencing problem has been carried out, are simply modelled by the following objective function (see $g(v)$ and its characteristic curve in Figure 2)

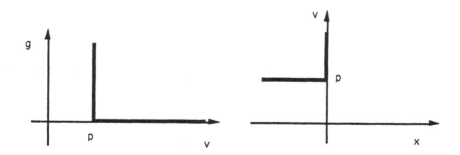

Figure 2

$$g_e(v_e) := \begin{cases} 0 & \text{if } v_e = u(T') - u(T) = S(T') - S(T) \ge p(T) \\ +\infty & \text{otherwise} \end{cases} \tag{23}$$

with $e = (T, T')$, for all T, T' which are consecutive in the linear ordering for some resource.
Precedence constraints can be modelled for any value of α and β as

$$g_e(v_e) := \begin{cases} 0 & \text{if } \alpha(T,T') + p(T) \leq v_e \leq \beta(T,T') + p(T) \\ +\infty & \text{otherwise} \end{cases} \tag{24}$$

with $e = (T, T')$.

We have also to model time constraints but these can be efficiently taken care of in the context of tardiness and earliness sum minimization. Therefore in order to model the sum of the weighted tardiness and earliness we build arcs (T^0, T) for all tasks T. On these arcs the objective functions are as follows (see $g(v)$ and its characteristic curve in Figure 3)

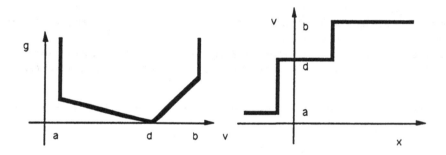

Figure 3

$$g_e(v_e) := \begin{cases} +\infty & \text{if } v_e < a(T) \\ \nu(T)\,(d(T) - v_e - p(T)) & \text{if } a(T) \leq v_e \leq d(T) - p(T) \\ \omega(T)\,(v_e + p(T) - d(T))) & \text{if } d(T) - p(T) \leq v_e \leq b(T) - p(T) \\ +\infty & \text{if } v_e > b(T) - p(T) \end{cases} \tag{25}$$

with $e = (T^0, T)$ and $v_e = u(T) - u(T^0) = S(T)$.

In order to model the maximum tardiness and maximum earliness we build arcs (T, T^+) and (T, T^-) for all tasks T plus two arcs (T^0, T^+) and (T^0, T^-) with functions:

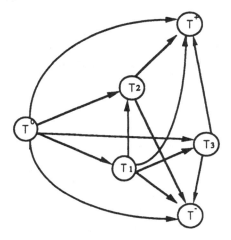

Figure 4

$$g_e(v_e) := \begin{cases} 0 & \text{if } v_e = u(T^+) - u(T) \geq p(T) - d(T) \\ +\infty & \text{otherwise} \end{cases} \quad \text{with } e = (T, T^+)$$

$$g_e(v_e) := \begin{cases} \omega_{max} v_e & \text{if } v_e = u(T^+) - u(T^0) \geq 0 \\ +\infty & \text{otherwise} \end{cases} \quad \text{with } e = (T^0, T^+)$$

$$g_e(v_e) := \begin{cases} 0 & \text{if } v_e = u(T^-) - u(T) \leq p(T) - d(T) \\ +\infty & \text{otherwise} \end{cases} \quad \text{with } e = (T, T^-)$$

$$g_e(v_e) := \begin{cases} -\nu_{max} v_e & \text{if } v_e = u(T^-) - u(T^0) \leq 0 \\ +\infty & \text{otherwise} \end{cases} \quad \text{with } e = (T^0, T^-)$$

$$(26)$$

In Figure 4 a very simple example is provided with three tasks T_1, T_2 and T_3; T_1 and T_2 share a resource, T_1 and T_3 share another resource and T_2 and T_3 have no resource in common so that there are the disjunctive arcs $T_1 \rightarrow T_2$ and $T_1 \rightarrow T_3$. Let us suppose that the optimal schedule is such that T_1 is completed before its due date, T_3 is completed after its due date and T_2 is completed exactly at its due date. Then the following flows are optimal: $x(T^0, T^+) := \omega_{max}$, $x(T^0, T^-) := -\nu_{max}$, $x(T^0, T_1) := -\nu(T_1)$, $x(T^0, T_2) := \nu_{max} + \nu(T_1) - \omega_{max} - \omega(T_3)$, $x(T^0, T_3) := \omega(T_3)$, $x(T_1, T^-) := \omega_{max} - \nu(T_1) + \omega(T_3)$, $x(T_2, T^-) := x(T^0, T_2)$, $x(T_3, T^+) := -\omega_{max}$, $x(T_1, T_3) := -\omega_{max} - \omega(T_3)$, all other flows being null provided the weights are such that $\nu_{max} \geq \omega_{max} + \omega(T_3) - \nu(T_1) \geq 0$ and $-\nu(T_2) \leq x(T^0, T_2) \leq \omega(T_2)$ hold.

This way a minimum cost circulation problem has been built, whose minimization corresponds to minimizing the following objective function

$$\nu_{max} \max_T V(T) + \omega_{max} \max_T W(T) + \nu_{sum} V_{sum} + \omega_{sum} W_{sum}.$$

Let us remark that the objective function (2) considers maximum weighted tardiness and earliness. If we want to minimize (2) a more complex mathematical programming problem should be solved. First we should drop from the network the nodes T^+ and T^- together with all incident arcs (this removes from the objective function all terms referring to the maximum earliness and tardiness). Then we should introduce the following bounds on the tensions of all arcs (T^0, T):

$$-\frac{V}{\nu(T)} + d(T) - p(T) \leq v_{(T^0,T)} = S(T) \leq \frac{W}{\omega(T)} + d(T) - p(T) \qquad \forall T.$$

These bounds could be viewed as given by a convex function $h : E \to \mathcal{R}^2$ constrained by $h(v) \leq (V, W)$. So we might consider the following convex mathematical programming problem:

$$G(V, W) := \inf \quad \sum_e g_e(v_e)$$
$$h(v) \leq (V, W)$$

It is well known that $G(V, W)$ is convex and so computing $\min_{(V,W)} G(V, W) + \nu_{max} V + \omega_{max} W$ can be done via standard techniques.

However we do not follow this approach. First maximum weighted earliness and tardiness have been taken care of by Problem $\mathcal{E}$. Secondly we have to reflect that assessing weights to the tasks is a subjective operation which is motivated by the wish to attach more importance to the completion time of some task with respect to other tasks. So there is no strict enforcement that this should be done in only one way, for instance by weighting the tardiness of each task and considering the maximum weighted tardiness. Alternatively we may also attach different importance to the different tasks by considering at the same time the maximum unweighted tardiness plus the weighted tardiness sum as it is embodied in the above described network flow model.

In order to solve the minimum cost network flow model several alternative approaches are available. The characteristic curve of each arc suggests using the out-of-kilter method. Although this is a fast method we preferred to employ a polynomial algorithm. For the same reason we did not consider using the network flow simplex algorithm. We decided to implement the algorithm developed by Goldberg and Tarjan (1987,1990) which uses very sophisticated data structures.

One important remark should be made at this point. If the objective function depends only on the tardiness of each task in a monotone way, then the schedule associated with the sequencing problem, if computed according to the earliest starting times, is already optimal for the scheduling problem. Similarly if the objective function depends only on the earliness of each task in a monotone way, then the schedule associated with the sequencing problem, if computed according to the latest starting times, is already optimal for the scheduling problem. In these extreme cases the reason for solving the scheduling problem is that it provides information about the critical set of tasks for any type of objective function. More details about the identification of the critical set of tasks are given in the next section.

There is another point concerning the scheduling problem which deserves some attention. We may think of extending the problem as defined in the IIASA exercise by giving the possibility of reducing the processing time for some task and correspondingly increasing some cost in term of money (the so called 'crashing'). This can be naturally taken care of in the context of the network flow problem. It is simply matter of adding one node T_c, representing the completion of the task T, for all tasks which can be possibly squeezed, assigning the following cost function to the arc (T, T_c)

$$g_e(v_e) := \begin{cases} +\infty & \text{if } v_e < p_s(T) \\ K(T)(p(T) - v_e) & \text{if } p_s(T) \le v_e \le p(T) \\ +\infty & \text{if } v_e > p(T) \end{cases}$$

where $K(T)$ is an assigned crashing cost, and replacing all arcs outgoing from T with arcs outgoing from T_c whose cost functions are like (23)-(26) with the term $p(T)$ dropped wherever it appears.

With this extension the objective function is made up of two parts, the first one measuring the cost in term of time and the second one measuring the cost in term of money. The parameter K can be used to assess the weight in the linear combination of the two cost components. By suitably changing K the entire Pareto optimal solutions with respect to the two objectives of money and time may be displayed.

9. THE IDENTIFICATION OF THE CRITICAL TASKS

In this section we investigate the problem of defining a set of critical tasks on the basis of the current schedule, which may have been provided either by Problem $\mathcal{E}$ or Problem $\mathcal{S}$. The tasks can be selected manually or on the basis of some automatic heuristic rule. We first discuss some criteria which may guide a manual selection and then an automatic procedure.

In order to understand the problem characteristics let us first note the difference between conjuctive and disjunctive arcs. The conjunctive arcs constitute intrinsic constraints of the problem and therefore the task sequencing induced by them is fixed independently of the schedule. A conjunctive arc defines a separation in time between two tasks and their predecessor and successor classes respectively, in such a way that there is no conflict in assigning the same resource to tasks in different classes. This suggests reassigments of resources restricted to either class of tasks. On the contrary disjunctive arcs can be removed by a different resource assignment, which avoids conflicts, thus providing a possibly improved schedule. In both cases (conjunctive and disjunctive arcs) the reassignment of resources cannot be done without taking into account all tasks processed during the same period of time, or *time window*; otherwise the improvement for some tasks could be obtained at the expense of other tasks. Once a time window has been identified by means of one of the presented criteria, a new assignment problem $\mathcal{A}(\mathbf{T}')$ will be solved restricted to the set of tasks $\mathbf{T}'$ which are scheduled in a time interval that overlaps with the time window.

Let us discuss in detail a number of criteria for the selection of time windows on the basis of the Gantt representation of a schedule.

1) Take a conjunctive arc on a critical path and consider as time window the time interval during which one of the two tasks involved is scheduled.

This selection criterion is motivated by the previous discussion. Consider the following example.

Example 1: A set of tasks $\mathbf{T}$ is given which consists of two tasks T and S, that is $\mathbf{T} = \{T, S\}$. A precedence constraint between the tasks is defined so that $T \prec S$. The resource set is $\mathbf{R} = \{R_1, R_2\}$. The modes of task T are such that $F_1^1 = \{R_1\}$ with $p(T, F_1^1) = 10$ and $F_2^1 = \{R_1, R_2\}$ with $p(T, F_2^1) = 6$. For what concerns task S, $F_1^2 = \{R_2\}$ with $p(S, F_1^2) = 10$ and $F_2^2 = \{R_1, R_2\}$ with $p(S, F_2^2) = 6$. The makespan has to be minimized.

In this simple case, the optimal schedule can be immediately obtained by $A(T) = F_1^1$ and $A(S) = F_1^2$. The solution of Problem $\mathcal{A}(\mathbf{T})$ gives the assignment $A(T) = F_1^1$ and $A(S) = F_1^2$ with $y_M = 10$. However, due to the presence of the conjunctive arc, the solution of Problem $\mathcal{E}$ provides a schedule with makespan $t_M(A) = 20$. On the basis of criterion 1) the time window [0,10] is selected and problem $\mathcal{A}(\{T\})$ is solved providing $A(T) = F_2^1$. The new schedule is such that $t_M(F) = 16$. A subsequent application of criterion 1) identifies the time window [6,16], and the solution of $\mathcal{A}(\{S\})$ provides $A(S) = F_2^2$ and the new schedule, with $t_M(A) = 12$, is optimal.

2) Take a chain of disjunctive arcs on a critical path, that link tasks $T_1, \ldots, T_k$ such that

$$\bigcap_{i=1,\ldots,k} A(T_i) \neq \emptyset.$$

The time window is identified as the time interval during which the tasks on the chain are scheduled, that is $[t(T_1), t(T_k) + p(T_k, A(T_k))]$.

In general such a chain reveals a critical time period for a resource, so that a better resource distribution in the time window identified through criterion 2) can improve the resulting schedule. The following example should clarify the concept.

Example 2: Let us consider a set of tasks $\mathbf{T} = \{T, S, U\}$ among which no precedence constraint is settled. The resource set is $\mathbf{R} = \{R_1, R_2, R_3\}$. A single mode is given for tasks T and S, such that $F_1^1 = \{R_1, R_2\}$ and $F_1^2 = \{R_1, R_3\}$ with $p(T, F_1^1) = p(S, F_1^2) = 10$. Two modes are given for task U, such that $F_1^3 = \{R_2, R_3\}$ with $p(U, F_1^3) = 10$ and $F_2^3 = \{R_2\}$ with $p(U, F_2^3) = 15$. The makespan has to be minimized.

Problem $\mathcal{A}(\mathbf{T})$ preferes for task U the faster mode F_1^3 obtaining $y_M = 20$, but the makespan obtained through problem $\mathcal{E}$ is $t_M(A) = 30$, as the tasks have to be sequenced. As the time window [0,20] identifies the tasks T and S for which a single mode is given, let us consider the time window [10,30] selected through application of criterion 2). Solution of a new assignment problem provides $A(U) = F_2^3$ with $y_M = 15$. Then the optimal schedule is obtained with $t_M^* = 25$.

Example 2 shows a case in which the assignment problem $\mathcal{A}(\mathbf{T})$ fails to find the optimal solution because of the resource set structure and criterion 2) can be successfully applied. We present now an example in which the failure is due to the presence of a conjunctive arc and criterion 2) can be still successfully applied.

Example 3: A set of tasks $\mathbf{T} = \{T, S, U\}$ is given with the following precedence constraints: $T \prec S$ and $T \prec U$. The resource set is $\mathbf{R} = \{R_1, R_2\}$. For what concerns task T, $F_1^1 = \{R_1\}$ with $p(T, F_1^1) = 20$. Two modes are given for both tasks S and U. The first mode for both tasks is such that $F_1^2 = F_1^3 = \{R_2\}$ with $p(S, F_1^2) = p(U, F_1^3) = 10$, while the second mode is such that $F_2^2 = F_2^3 = \{R_1\}$ with $p(S, F_2^2) = p(U, F_2^3) = 1$. Still the makespan has to be minimized.

Problem $\mathcal{A}(\mathbf{T})$ distributes resources as uniformly as possible among all tasks providing $A(T) = F_1^1$, $A(S) = F_1^2$ and $A(U) = F_1^3$. Solution of $\mathcal{E}$ introduces an arc between tasks S and U so that $t_M(A) = 40$ which is not optimal. Criterion 2) allows the identification of the time window [20,40] so that solution of problem $\mathcal{A}(\{S, U\})$ provides the optimal assignment with $A(S) = F_2^2$ and $A(U) = F_2^3$. The optimal schedule has makespan $t_M^* = 22$.

Another criterion which may be sensibly considered in the selection of time windows consists in taking into account release dates, due dates and deadlines, as already anticipated. There are many possible ways to exploit this information for the selection of time windows. One possibility consists in selecting as time window any slot given by two successive dates, no matter whether they are release dates, due dates or deadlines.

As already mentioned, the decision about the time window selection can be left to the decision maker, and it is therefore important to provide the decision maker with some tools in order to enhance his decisions. We have limited ourselves to display the critical path in a graphical form by using typical GANTT charts.

Now we deal with the problem of identifying the critical tasks by using the solution of Problem $\mathcal{S}$. We recall that the solution of Problem $\mathcal{S}$ consists of a set of dual variables, i.e. the potentials corresponding to the starting times of the tasks, and a set of primal variables assigning flow values to the arcs. Being these flows dual variables of the original problem we may interpret them as quantities measuring the sensitivity of the objective with respect to the problem data.

In particular for all arcs e whose function $g_e(v_e)$ is of the type

$$g_e(v_e) := \begin{cases} 0 & \text{if } -\infty \leq a \leq v_e \leq b \leq +\infty \\ +\infty & \text{otherwise} \end{cases}$$

the variation rate $\Delta\psi$ for the objective function is given by $\Delta\psi = -x_e\Delta a$ if $v_e = a$ and $\Delta\psi = x_e\Delta b$ if $v_e = b$. So we may consider as critical the arcs carrying large flows, since a tighter schedule for the relative tasks could lead to a larger decrease for the objective function.

In the extreme case there is no feasible solution to the scheduling problem, the flow problem is unbounded. In this case a cycle can be detected where an unbounded flow may circulate. Therefore if we want to revise the assignment (or the seqences) in order to have a feasible schedule we have a direct information about which tasks should receive a different assignment and which resources should be resequenced.

10. DESCRIPTION OF THE DSS

Following Silver (1988) we describe the DSS through its functional capabilities, the user view of system components and the system attributes.

Functional capabilities - The functional capabilities answer the question "what can the system do?" and have been made clear in the description of the problem itself. Therefore we will not discuss them any longer.

User view of system components - We start describing the user view of system components, which answers the question "what does the system look like?", describing the operators offered. The main operators, corresponding to the main menu are: File, Edit, Check, Schedule, Display, Exit.

With the File operator, a data file already existing can be opened or a new file can be created. The Edit operator allows an editing of the data file, that is the modification of an existing data file or the generation of a new one. Obviously the data file can be generated outside the DSS with a general purpose editor and then only opened inside the DSS with the File operator. The Check operator checks the formal correctness of the data file. First of all, the format of the file must be as defined by the exercise and all the crossed references among the tables which define the problem instance must be correct. It is checked whether the precedence graph defined in the table *precedence* contains cycles. Moreover, some checking is made on the numerical values defined in the tables. In particular, for each pair of tasks (T, T') defined in the table *precedence* it must be

$$\alpha(T, T') \leq \beta(T, T').$$

The release dates and the deadlines of the tasks must be well defined, that is it is checked whether

$$a(T) \leq d(T) \leq b(T) \quad \forall T.$$

The availability intervals of each resource must be well defined disjoint intervals. A non-negativity check is made on the data which need it. In all cases in which an error is identified, an explanation message appears on the screen. The Schedule operator cannot be accessed if the data file has not been checked with the Check operator. The Schedule operator is in fact a list of operators and will be discussed later. The Display operator shows the problem instance by means of numerical tables. The Exit operator allows the user to quit the DSS.

The list of operators offered by Schedule is: File, Edit, Check, Solve, Display, Instance. Analogously with the operators previously defined for the data file, the File operator allows the loading of an already existing schedule file or the creation of a new one and the Edit operator allows the editing of the schedule file. The Check operator checks the feasibility of the schedule. The Solve operator deals with the generation of a schedule by means of automatic, interactive and manual procedures and will be discussed in more detail in the following. The Display operator shows the solution corresponding with the opened schedule file by means of task Gantt chart. The Instance operator allows going back to the instance data operators.

The Schedule operator makes available a number of sub-operators: Criteria, Manual, Interactive, Auto, Evaluate. The Criteria operator allows the selection of the criterion on the basis of which the schedules will be generated and evaluated. The Manual operator supports the manual generation of a schedule (this feature has not been implemented yet). The Interactive operator will be discussed shortly. The Auto operator automatically generates a schedule (at the moment it solves in sequence problems $\mathcal{A}$, $\mathcal{E}$ and $\mathcal{S}$). The Evaluate operator shows a table of values evaluating the quality of the schedule.

The Interactive operator makes available the sub-operators: Assignment, Sequencing, Scheduling, Window, Cycle. The Assignment operator allows the resource assignment by selecting one or more procedures among the ones described in Section 6. A new mode is selected for all the tasks defined in a current task set. When the Assignment operator is selected for the first time the current task set is defined as the global task set **T**. After a procedure has been selected, information on the value of the objective function obtained by the procedure and the resource which causes the maximum resource usage are shown. The Sequencing operator generates the sequencing of the tasks according to the procedure described in Section 7. During the execution of the procedure the screen shows detailed information about the evolution of the procedure.

The Scheduling operator finds a schedule for all tasks by means of the procedure described in Section 8. The Window operator allows the user to select a subset of tasks which becomes the current set of tasks, denoted as **T'** in Section 9. Finally, the Cycle operator allows the automatic execution of the operators: Assignment (on the current task set), Sequencing and Scheduling.

System attributes - In describing the system attributes the attention moves to the decision making process. The question to answer here is: "How will the system affect the decision making?". We will discuss how much the user is free or restricted in creating a schedule.

The different solution approaches, interactive, one cycle, automatic, manual are included in order to offer a good trade-off between creativity and support. The user can either prepare a manual schedule (in the present version of the DSS this option has still to be implemented) or to use the interactive procedure or to use the automatic procedure. The interactive procedure tries to capture the steps of the decision process, resource assignment, sequencing and scheduling, in order to provide the user with the algorithmic support while mantaining a natural way of generating a solution, making therefore easy and powerful his possibility of interacting with the system in the solution generation process. Infeasibility is not allowed. Many solutions can be saved. The solutions can be partially modified. Many procedures are offered for the resource assignment problem. Most of them are very fast so that the user can test many of them.

Assistance to the user is given by means of messages which appear on the bottom line of the screen to explain what the current state of the system does and offers. The user is not supported in the selection of the solution approach, as the difference among manual, interactive and automatic should be clear enough. Among the procedures offered for the resource assignment problem, the only reason for selecting one instead of another is the computational time. Therefore if the option "simulated annealing" or "branch-and-bound" are chosen a message appears which warns the user about the risk of a high processing time.

However, the user can interrupt the procedures at any time obtaining the best solution up to that moment.

The possibility of creating several schedules gives the user the opportunity to a posteriori analyse the solutions, to see the Gantt charts, to print them, to evaluate them on the basis of different objectives. If he assumes that one identified objective perfectly represents his decision criterion, he can simply select the solution which optimizes such objective, otherwise he should select on the basis of a personal analysis of the solutions.

11. COMPUTATIONAL RESULTS

We have designed the DSS with the goal of obtaining good schedules with a small effort both in terms of computing time and human interaction. Therefore our tests (on a IBM/AT, as required by the exercise) have been first directed to verify the speed of the basic algorithms, that is the assignment algorithms described in Section 6, the Bottleneck Shifting Procedure and the One Machine Problems described in Section 7 and finally the circulation algorithm of Section 8. Since the last algorithm is polynomial and produces the optimal solution, there is actually no need of a thourough testing procedure for it.

Moreover there is the need to measure the quality of the solution provided by the heuristics with respect to the objective function of the problem. This is a crucial point since no optimal solution is in general available for the subproblems and a comparison can be made only with respect to other heuristic algorithms. However, two instances are available for which the optimal solution is known. These are the famous 6×6 and 10×10 Job Shop Problems proposed by Fisher and Thompson (1963) which constitute classical benchmark problems for Job Shop Scheduling. Actually our problem is different in many respects from a simple Job Shop and so this comparison should be judged with care.

Assignment algorithms - As far as the speed is concerned the greedy heuristics behave satisfactorily by providing a solution in less than one second. The branch and bound algorithm is obviously much slower leading sometimes to unacceptable computing times of several minutes. Therefore the use of the branch and bound heuristic is mandatory. It has been observed that this heuristic finds the optimal solution in most cases and that finds a solution very close to the optimum in the remaining cases. Still this heuristic is not very fast, half a minute may be necessary for the largest instances with one hundred tasks. One reason is that the code has not been optimized. Actually a full linear programming problem is solved at each node of the search tree without exploiting the optimal basis of the father node as it is usually done in all branch and bound codes based on linear programming. It should be anyway remarked that this improvement can be done at the expense of memory size since the optimal basis matrix should be stored for each generated subproblem.

Since there is no actual need of an optimal solution for the assignment problem being this a rather crude approximation of the full problem, we think that, at least in the first step, when the problem is large, it is sensible to use the fast greedy heuristics. However, in a subsequent phase of the iterative procedure when only a small subset of the tasks have to be reassigned and typically these tasks interact in a time window where there are few conjunctive precedences among them, the assignment model approximates better the general problem and finding an optimal solution can be both useful and fast.

Sequencing algorithms - The Shifting Bottleneck Procedure is the real core of the entire DSS. We have devoted a great care for its design and the results we have got can be considered very satisfactory. First of all we mention the results concerning the 6×6 and 10×10 Job Shop Problems. For the smaller problem the optimal solution is found in less than one second. For the larger problem a solution with value 950 for the makespan is found in 8 seconds. We recall that the optimal makespan is 931 and existing algorithms find it after a very long processing time (one hour cpu time to find it and four more hours to prove its optimality, see Carlier and Pinson, 1986).

In order to test our procedure on instances typical of the problem we are dealing with, we have followed two approaches: first we have generated random instances, second we have modified the two 6×6 and 10×10 instances. The first approach is typical for testing procedures and so we have generated three sets of random instances with 25, 50 and 100 tasks fixing the number of resources to 5, 10 and 20 respectively with at most three resources per task. The conjunctive precedence graph has been generated with at most three successors per task. The processing times for the tasks have been uniformly generated between 1 and 30. The release dates have been generated randomly and then they have been possibly corrected by taking into account the structure of the precedences and the processing times of the tasks. The same approach has been adopted for the deadlines by also imposing that the minimum interval between the release date and the deadline for a task is five times the processing time. The due date has been generated randomly within this interval. We have only tested the procedure with respect to maximum tardiness minimization and all tasks equally weighted. This actually corresponds to the most critical combinatorial situation. Furthermore a certain number of unavailability intervals for the resources have been generated.

Our heuristic has been tested against a bunch of fast heuristic like, FIFO (First In First Out), LIFO (Last In First Out), SPT (Shortest Processing Time), LPT (Longest Processing Time), NDL (Next DeadLine), MNS (Most Number of Successors), Random. In these heuristics one queue of tasks and one queue of resources are formed. At each instant of time a task is selected according to the priority rule specified by the heuristic if the corresponding resources are in the resource queue. Then the task and the resources are pushed from the queues and the resources will reenter the resource queue at the task completion time. Of course no task can be selected if it has to be processed during some unavailability interval concerning its resources.

We have observed that our heuristic produces solutions whose maximum tardiness is better than the solution provided by the fast heuristics by an amount which is on the average two or three times the average processing time. Of course these heuristics are much faster even if taken as a whole. This result is not entirely in favour of our heuristic since in our opinion the improvement in solution quality does not pay the poorer computing time behaviour, not to say of the simplicity in devising codes for the fast heuristics.

Actually we have found some difficulty in generating instances exhibiting a strong combinatorial structure. Indeed the randomly generated instances turn out to be 'simple' solutions, due to the presence of a certain number of conjunctive precedences and a lot of due dates which, in some sense, already suggest how the solution should look like.

Following these considerations we have decided to generate two instances directly from the 6×6 and 10×10 instance solutions, by adding some features typical of our problem. More precisely we have specified the following resource assignments for the tasks ($T_{i,j}$ is the j-th task of the i-th job): $A(T_{1,2}) := \{R_1, R_6\}$, $A(T_{1,5}) := \{R_3, R_4, R_6\}$, $A(T_{2,2}) := \{R_3, R_5\}$, $A(T_{2,4}) := \{R_2, R_6\}$, $A(T_{3,5}) := \{R_2, R_3\}$, $A(T_{5,2}) := \{R_2, R_4\}$, $A(T_{6,1}) := \{R_2, R_4\}$, $A(T_{6,6}) := \{R_2, R_3\}$, all other tasks being assigned to one single resource as in Muth and Thompson?. The resource R_1 is declared unavailable during [9,13] and [27,28]; R_3 during [30,35] and R_2 during [40,47] all other resources being always available. The following deadlines have been assigned: $b(T_{1,2}) := 9$, $b(T_{1,3}) := 22$, $b(T_{1,5}) := 41$, $b(T_{3,3}) := 17$, $b(T_{3,4}) := 27$, $b(T_{3,5}) := 28$, $b(T_{5,5}) := 51$, $b(T_{6,2}) := 19$, $b(T_{6,5}) := 49$, all other tasks being assigned very large deadlines. Finally release and due dates $a(T) = d(T) := 0$ have been assigned for all tasks.

We remark that these modifications have been carried out so that the optimal solution of the corresponding job shop problem is still feasible and optimal for the new data. Our procedure yields the same optimal solution with makespan value 55 (it coincides with the maximum tardiness sind $d(T) = 0$) within two seconds computing time. The fast heuristics provide a feasible solution with makespan 71.

We have carried out a similar modification for the 10×10 instance. These are the modified resource assignments: $A(T_{1,6}) := \{R_6, R_7\}$, $A(T_{2,7}) := \{R_1, R_3, R_7\}$, $A(T_{3,7}) := \{R_2, R_3, R_6, R_8\}$, $A(T_{4,2}) := \{R_3, R_8, R_{10}\}$, $A(T_{5,1}) := \{R_3, R_4, R_6, R_7\}$, $A(T_{5,2}) := \{R_1, R_6\}$ $A(T_{6,5}) := \{R_4, R_8, R_9, R_{10}\}$, $A(T_{6,9}) := \{R_3, R_5\}$, $A(T_{8,1}) := \{R_3, R_6\}$, $A(T_{8,6}) := \{R_6, R_7\}$, $A(T_{8,10}) := \{R_2, R_4\}$, $A(T_{9,1}) := \{R_1, R_5, R_7, R_9\}$, $A(T_{9,9}) := \{R_1, R_2, R_5\}$. The resource R_1 is unavailable during [211-224], R_2 during [600,620], R_4 during [370-380], R_5 during [180-280], R_9 during [780-800], R_{10} during [520-531] all other resources being always available.

These data are such that the solution found for the job shop problem with makespan 950 is still feasible (and presumably optimal) for the new instance. However, our procedure is not able to find it. It yields a solution with makespan 1132 which is in any case much better than the makespan 1350 given by the fast heuristcs. The required computing time for our procedure is 20 seconds.

12. CONCLUSIONS

A multi-stage decomposition approach for the resource constrained scheduling problem has been presented. It provides a powerful environment for combining algorithmic support and human interaction in order to find satisfactory solutions. Each of the resulting subproblems has been carefully formulated and analyzed. The protoype of the DSS embeds all the subproblems and the algorithms described. Future efforts will be devoted to the improvement of the prototype, in particular the interactive parts and the graphical representation of instances and of partial and global solutions.

13. ACKNOWLEDGMENTS

The authors acknowledge the contribution of the students which have worked in the implementation of the prototype and in particular the contribution of Giuseppe Lancia.

14. REFERENCES

ADAMS, J., E. BALAS AND D. ZAWACK, 1988, "The Shifting Bottleneck Procedure for Job Shop Scheduling", *Management Science*, 34, 391-401

ANTHONISSE, J.M., K. M. VAN HEE AND J.K. LENSTRA, 1988, "Resource-Constrained Project Scheduling: an International Exercise in DSS Development", *Decision Support Systems*, 4, ?

CARLIER, J., 1982, "The One-Machine Sequencing Problem", *European J. of Operational Research*, 11, 42-47

CARLIER, J. AND E. PINSON. 1986, "Une méthode arborescente pour optimiser la durée d'un job-shop", Le cahiers de l'Institut de Mathématiques Appliquées, Angers, ISSN 0294-2755

CHRISTOFIDES, N., R. ALVAREZ-VALDES AND J.M. TAMARIT, 1987, "Project Scheduling under Resource Constraints", *European J. of Operational Research*, 29, 262-273

FISHER, H. AND G.L. THOMPSON, 1963, "Probabilistic learning. Combinations of local Job Shop Scheduling Rules", *Industrial Scheduling*, J.F. Muth and G.L. Thompson, eds., Prentice Hall, Englewood Cliffs, N.J., 225-251

GAREY, M.R. AND D.S. JOHNSON, 1979, *Computers and Intractability: a Guide to the Theory of NP-completeness*, Freeman, San Francisco

GOLDBERG, A.V. AND R.E. TARJAN, 1987, "Solving Minimum-Cost Flow Problems by Successive Approximation", *Proc. 19th Annual ACM Symposium on Theory of Computing*, 7-18

GOLDBERG, A.V. AND R.E. TARJAN, 1990, "Finding Minimum Cost Circulations by Successive Approximation", *Mathematics of Operations Research*, 15, 430-466

LANCIA, G., 1990, "Combinatorial and Network Flow Algorithms for Optimal Project Scheduling", Thesis, Faculty of Sciences, University of Udine (in English)

LENSTRA J.K. AND A.H.G. RINNOOY KAN, 1985, "Sequencing and Scheduling", *Combinatorial Optimization: Annotated Bibliographies*, M. O'hEigeartaigh, J.K. Lenstra and A.H.G. Rinnooy Kan, eds., Wiley and Sons

PATTERSON, J.H., 1984, "A Comparison of Exact Approaches for Solving the Multiple Constrained Resource Project Scheduling Problem", *Management Science*, 30, 854-867

RINNOOY KAN, A.H.K., 1976, *Machine Scheduling Problems: Classification, Complexity and Computations*, Nijhoff, The Hague

ROCKAFELLAR, R. T., 1984, *Network Flows and Monotropic Optimization*, Wiley and Sons

SERAFINI, P. AND M.G. SPERANZA, 1988, "Resource Assignment in a DSS for Project scheduling", *Proceedings of the Conference "Multiobjective Problems of Mathematical Programming"*, Yalta October 26 – November 2 1988

SERAFINI, P. AND W. UKOVICH, 1988, "A Decision Support System Based on the Job Shop Problem in Real Manufacturing", *Methodology and Software for Interactive Decision Support*, IIASA ed., Springer, Berlin

SERAFINI, P., W. UKOVICH, H. KIRCHNER, F. GIARDINA AND F. TIOZZO, 1989, "Job Shop Scheduling: a Case Study", *Operations Research Models In Flexible Manufacturing Systems*, F. Archetti, M. Lucertini and P. Serafini, eds., Springer, Vienna

SILVER, M.S., 1988, "Descriptive Analysis for Computer-Based Decision Support", *Op. Res.*, **36**, 904-916

SPERANZA, M.G. AND C. VERCELLIS, 1990, "Hierarchical Models for Multi-Project Planning and Scheduling", *Proceedings of the "Second Workshop on Project Management and Scheduling"*, Compiegne, France, June 20–22, 1990

CONCEPTS OF THE REFERENCE POINT CLASS OF METHODS OF INTERACTIVE MULTIPLE OBJECTIVE PROGRAMMING

R.E. Steuer

University of Georgia, Athens, Georgia, USA

L.R. Gardiner

Auburn University, Auburn, Alabama, USA

This paper discusses topics fundamental to the understanding of the reference point methods of interactive multiple objective programming (represented by the work of Benayoun, de Montgolfier, Tergny and Larichev (1971), Wierzbicki (1977, 1982 and 1986), Steuer and Choo (1983), Nakayama and Sawaragi (1984), Korhonen and Laakso (1986), and others [6, 11, 12, 13, 14 and 18] that has attracted considerable attention in the 1980s.

The multiple objective program addressed is:

$$\max \ \{f_1(x) = z_1\}$$
$$\cdot$$
$$\cdot$$
$$\cdot$$
$$\max \ \{f_k(x) = z_k\}$$

$$\text{s.t.} \quad x \in S$$

where k is the number of objectives, the z_i are criterion values, and S is the feasible region in decision space. Let $Z \subset R^k$ be the feasible region in <u>criterion space</u> where z ϵ Z if and only if there exists an x ϵ S such

that z = $(f_1(x), \ldots, f_k(x))$. Let K = $\{1, \ldots, k\}$. <u>Criterion vector</u> $\bar{z}$ ϵ Z is

<u>nondominated</u> if and only if there does not exist another z ϵ Z such that z_i

$\geq \bar{z}_i$ for all i ϵ K and $z_i > \bar{z}_i$ for at least one i ϵ K. The set of all

nondominated criterion vectors is designated N and is called the

<u>nondominated set</u>. A point $\bar{x}$ ϵ S is <u>efficient</u> if and only if its criterion

vector $\bar{z} = (f_1(\bar{x}), \ldots, f_k(\bar{x}))$ is nondominated. The set of all efficient

points is designated E and is called the <u>efficient set</u>. Let U: $R^k \rightarrow R$ be a

<u>utility function</u> of a <u>decision maker</u> (DM). A $z° ϵ Z$ that maximizes U over

Z is an <u>optimal criterion vector</u> and any x° ϵ S such that

$(f_1(x°), \ldots, f_k(x°)) = z°$ is an <u>optimal solution</u> of the multiple objective

program. Our interest in the efficient set E and the nondominated set N

stems from the fact that if U is <u>coordinatewise increasing</u> (that is, "more

is always better than less" of each criterion), x° ϵ E and z° ϵ N.

Consider the multiple objective program

$$
\begin{aligned}
\max \ \{ &\text{profit} & = z_1\} \\
\max \ \{ &\text{product innovation} & = z_2\} \\
\max \ \{ &\text{manufacturing safety} & = z_3\} \\
\text{s.t.} & & z ϵ Z
\end{aligned}
$$

in which, for a criterion vector to be a <u>contender</u> for optimality, the

criterion vector must be nondominated. Let z^1, z^2, z^3 ϵ Z where:

$$
z^1 = \begin{bmatrix} 5 \\ 2 \\ 1 \end{bmatrix}
\qquad
z^2 = \begin{bmatrix} 2 \\ 7 \\ 3 \end{bmatrix}
\qquad
z^3 = \begin{bmatrix} 6 \\ 8 \\ 3 \end{bmatrix}
$$

With regard to the three criterion vectors, only a "fool" would choose

z^1 or z^2 over z^3 because z^1 and z^2 are <u>dominated</u> by z^3. Rather than

waste a DM's time with dominated criterion vectors (which cannot be
optimal), we pursue the strategy of searching the nondominated set N for
the DM's most preferred criterion vector. This criterion vector would
then be optimal for the multiple objective program. Unfortunately,
multiple objective programs typically have many nondominated criterion
vectors, and finding the best nondominated criterion vector is not a
trivial exercise.

One might think that the best way to solve a multiple objective
program would be to assess the DM's U and then solve the (single
objective) program

$$\max \; \{U(z_1, \ldots, z_k)\}$$
$$\text{s.t.} \quad f_i(x) = z_i \qquad i \in K$$
$$x \in S$$

because any solution that solves this program is an optimal solution of
the multiple objective program. However, multiple objective programs
are not solved in this way for several reasons. One is the difficulty
in obtaining an accurate enough representation of U for use in the above
program. Another is that even if it were possible to obtain a U, it
would almost certainly be nonlinear, causing the above to be a nonlinear
program. A third reason is that the above program produces only one
solution, and a DM may need to see several solutions before being
confident enough to accept any given solution as final.

Because of difficulty in locating the best nondominated criterion
vector, we usually conclude the search for an optimal solution with a
final solution, where a final solution is either optimal, or close
enough to being optimal to satisfactorily terminate the decision

process.

The paper is divided into three parts. Part 1 describes background topics necessary for understanding the reference point methods of interactive multiple objective programming. Part 2 covers topics related to the current status of reference point methods. Part 3 discusses future directions for reference point methods.

PART 1: BACKGROUND TOPICS

1.1   Decision Space Versus Criterion Space

Whereas the study of conventional (single objective) mathematical programming is conducted in decision space, interactive multiple objective programming is mostly studied in criterion space. To illustrate the difference between a multiple objective program's representation in decision space and its representation in criterion space, consider

$$\max \ \{x_1 - 1/2x_2 = z_1\}$$
$$\max \ \{ \qquad x_2 = z_2\}$$
$$\text{s.t.} \qquad x \in S$$

where S in decision space is given in Figure 1, and Z in criterion space is given in Figure 2. For instance z^4, which is the image of $x^4 = (3,4)$, is obtained by plugging (3,4) into the objective functions to generate $z^4 = (1,4)$. In this way, Z is the image of S under the f_i. In Figure 2, the nondominated set N is the set of boundary criterion vectors z^3 to z^5 to z^6. In Figure 1, the efficient set E is the set of inverse images of the criterion vectors in N, namely the set of boundary points x^3 to x^5 to x^6. The inverse image of a criterion vector tells us what must be done to achieve the results specified by the criterion vector. Note that Z is not necessarily confined to the nonnegative orthant.

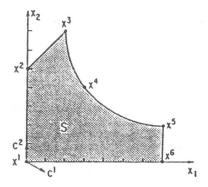

Figure 1. Representation in Decision Space.

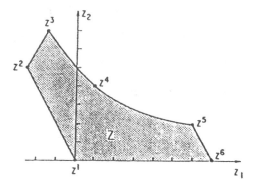

Figure 2. Representation in Criterion Space.

1.2 Unsupported Nondominated Criterion Vectors

A nondominated criterion vector $\bar{z} \in Z$ is <u>unsupported</u> if and only if it is not possible to support Z with a hyperplane at $\bar{z}$. In Figure 2, the set of unsupported nondominated criterion vectors is the set of criterion vectors from z^3 through z^4 to z^5, exclusive of z^3 and z^5. The set of <u>supported</u> nondominated criterion vectors is the set that consists of z^3 plus the line segment z^5 to z^6, inclusive. Unsupported nondominated criterion vectors can only occur in non-convex feasible regions, hence they can only occur in integer and nonlinear multiple objective programs. Reference point methods are designed not to be thwarted by unsupported criterion vectors.

1.3   Graphical Detection of Nondominated Criterion Vectors

Let $\bar{z} \in Z$. Let $D_{\bar{z}}$ be the set formed by translating the nonnegative orthant of R^k to $\bar{z}$. $D_{\bar{z}}$ then shows the points in R^k that dominate $\bar{z}$. If, except for $\bar{z}$, none of the points in $D_{\bar{z}}$ are feasible, $\bar{z}$ is nondominated. In other words, $\bar{z}$ is a nondominated criterion vector if and only if $Z \cap D_{\bar{z}} = \{\bar{z}\}$. Visualizing in Figure 2 the nonnegative orthant translated to z^4, we see that z^4 is nondominated. Visualizing the nonnegative orthant translated to z^2, we see that z^2 is dominated (because, for instance, $z^3 \in Z$ is in D_{z^2}).

1.4   Payoff Tables

A <u>payoff table</u> is of the form

	z_1	z_2		z_k
z^1	z_1^*	z_{12}		z_{1k}
z^2	z_{21}	z_2^*		z_{2k}
			.	
			.	
			.	
z^k	z_{k1}	z_{k2}		z_k^*

where the rows are criterion vectors resulting from individually maximizing the objectives. If an objective has alternative optima, a way to assure that the criterion vector of that row of the payoff table is nondominated is to <u>lexicographically maximize</u> the objectives.

The z_i^* entries along the <u>main diagonal</u> of the payoff table are the <u>maximal criterion values</u> for the different objectives over the nondominated set. The <u>minimum value in the i-th column</u> of the payoff table is an <u>estimate</u> of the <u>minimum criterion value of the i-th objective</u> <u>over N</u>. Often these column minimums are used in place of the minimum criterion values over N because the minimum criterion values over N are difficult to obtain (see Isermann and Steuer (1988)).

1.5 z* and z** Reference Criterion Vectors

The $z^* \in R^k$ <u>reference criterion vector</u> is formed from the z_i^* entries along the main diagonal of the payoff table where:

$$z_i^* = \max \{f_i(x) \mid x \in S\}$$

The $z^{**} \in R^k$ <u>reference criterion vector</u> is an infeasible criterion vector that strictly dominates every nondominated criterion vector $z \in$ N. The components of z^{**} are given by:

$$z_i^{**} = z_i^* + \epsilon_i$$

where the ϵ_i are moderately small positive values. An ϵ_i value that raises z_i^{**} to the smallest integer greater than z_i^* is normally sufficient.

1.6   Aspiration Criterion Vectors

An <u>aspiration criterion vector</u> $q \in R^k$ is a criterion vector
specified by a DM to reflect his or her hopes or expectations from the
model. Aspiration criterion vectors are usually specified in the light
of a payoff table such that $q \leq z^*$. Aspiration criterion vectors are
typically specified so that on the next iteration the solution procedure
will either (a) <u>project</u> q onto N (compute the nondominated criterion
vector closest to q), or (b) compute nondominated criterion vectors from
a region in N closest to q.

1.7   Range Equalization Weights

Suppose we have an objective such as profit that is measured in
millions, and another such as safety that is measured on a scale of 0 to
10. In order to prevent the objectives with the largest criterion
values from biasing the search for a final solution, it is never a bad
idea to normalize the objective functions. Let R_i be the <u>range</u> of z_i
over N (the difference between z_i^* and the minimum criterion value of the
i-th objective over N). A good way to normalize the objectives is to
equalize the R_i's by multiplying each objective by its π_i <u>range</u>
<u>equalization weight</u>

$$\pi_i = \begin{cases} 0 & \text{if } R_i = 0 \\[2ex] \dfrac{1}{R_i} \left[\sum_{j=1}^{k} \dfrac{1}{R_j} \right]^{-1} & \text{if } R_i > 0 \end{cases}$$

Because the minimum criterion values of the different objectives are
typically unavailable, each R_i can be replaced by the difference between

z_i^* and the minimum value in the i-th column of the payoff table, roughly achieving the same purpose.

1.8 Lexicographic and Augmented Weighted Tchebycheff Sampling Programs

Using a z** reference criterion vector, we have the <u>lexicographic weighted Tchebycheff sampling program</u> for generating points from the nondominated set

$$\text{lex min } \{\alpha, \ -\sum_{i=1}^{k} z_i)\}$$

$$\text{s.t.} \quad \alpha \geq \lambda_i(z_i^{**} - z_i) \qquad i \in K$$

$$f_i(x) = z_i \qquad i \in K$$

$$x \in S$$

$$\alpha \in R \text{ unrestricted, } z \in R^k \text{ unrestricted}$$

where $\lambda \in \Lambda = \{\lambda \in R^k | \lambda_i \in (0,1), \ \sum_{i=1}^{k} \lambda_i = 1\}$. The lexicographic sampling program is valuable because:

(1) If $\bar{z} \in Z$ is a minimizing criterion vector solution of the lexicographic sampling program, $\bar{z}$ is nondominated.

(2) If $\bar{z} \in Z$ is nondominated, there exists a $\lambda \in \Lambda$ such that $\bar{z}$ <u>uniquely</u> solves the lexicographic sampling program.

Result (1) states that the lexicographic sampling program will only generate nondominated criterion vectors. Result (2) means that no nondominated criterion vector can <u>hide</u> from the lexicographic sampling program.

Often used in place of the lexicographic sampling program is the

augmented weighted Tchebycheff sampling program

$$\min \{\alpha - \rho \sum_{i=1}^{k} z_i\}$$

$$\text{s.t. } \alpha \geq \lambda_i(z_i^{**} - z_i) \qquad i \in K$$

$$f_i(x) = z_i \qquad\qquad i \in K$$

$$x \in S$$

$\alpha \in R$ unrestricted, $z \in R^k$ unrestricted

where ρ is a sufficiently small positive scalar. The advantage of the

augmented sampling program is that it avoids lexicographic

optimizations, but its disadvantage is the sufficiently small scalar ρ.

1.9   Contours of the Lexicographic and Augmented Weighted Tchebycheff
Sampling Programs

Let the β level set of a function $g: R^k \to R$ be given by $\{z \in$

$R^k | g(z) \geq \beta\}$. Then the level sets in R^k of the objective function of

the first optimization stage of the lexicographic weighted Tchebycheff

sampling program are translated nonnegative orthants whose vertices lie

along the line that goes through z^{**} in the direction

$$(1/\lambda_1, \ 1/\lambda_2, \ \ldots, \ 1/\lambda_k)$$

In this way, the Tchebycheff contours (sets of points of fixed value) of

the objective function of the first optimization stage are "L-shaped"

sets whose "vertices" lie along the same line through z^{**}. Thus it can

be visualized that the first optimization stage determines the

Tchebycheff (L-shaped) contour furthest along the line that goes through z** in the direction given above. If there is only one feasible criterion vector on this Tchebycheff contour, this is the nondominated criterion vector returned by the lexicographic sampling program. If there is more than one feasible criterion vector on this Tchebycheff contour, some may be dominated. In this case, the second optimization stage is employed to determine a feasible criterion vector on the Tchebycheff contour that is closest to z** according to the L_1-metric. Such a criterion vector is nondominated.

The augmented Tchebycheff contours of the objective function of the augmented weighted Tchebycheff sampling program are as shown in Steuer (1986, Figures 14.4 and 15.9).

1.10 Determining Tchebycheff Vertex λ-Vectors

Consider a $\bar{z} \in R^k$ such that $\bar{z} < z^{**}$. Then, the Tchebycheff vertex λ-vector defined by $\bar{z}$ and z** is the λ-vector that causes the vertices of the Tchebycheff contours of the first optimization stage of the lexicographic weighted Tchebycheff sampling program (or the vertices of the augmented Tchebycheff contours of the augmented weighted Tchebycheff sampling program) to lie along the line that goes through $\bar{z}$ and z**. The components of the Tchebycheff vertex λ-vector are given by:

$$\lambda_i = \frac{1}{z_i^{**} - \bar{z}_i} \left[\sum_{j=1}^{k} \frac{1}{z_j^{**} - \bar{z}_j} \right]^{-1}$$

1.11 Groups of Evenly Dispersed λ-Vectors

Suppose we are using an interactive procedure that requires, on iteration h, a group of <u>evenly dispersed</u> λ-vectors from the following

<u>interval defined subset</u> of weighting vector space

$$\Lambda^{(h)} = \{\lambda \in R^k | \lambda_i \in (\ell_i^{(h)}, \mu_i^{(h)}), \sum_{i=1}^{k} \lambda_i = 1\}$$

$$\text{where } (\ell_i^{(h)}, \mu_i^{(h)}) \subset (0,1) \text{ for all } i \in K$$

so that the lexicographic or augmented weighted Tchebycheff sampling

programs can compute dispersed representatives from a neighborhood in N.

By evenly dispersed we mean λ-vectors that are as uniformly

distributed over $\Lambda^{(h)}$ as far as it is practicable to compute. Such

groups of dispersed λ-vectors can be obtained by using the LAMBDA and

FILTER codes from Steuer (1990). Suppose we need 10 dispersed λ-vectors

from $\Lambda^{(h)}$. Then the LAMBDA code would be used to produce an

overabundance of, say 200, randomly generated λ-vectors from $\Lambda^{(h)}$. Then

FILTER would select 10 λ-vectors from the 200 that are as far apart from

one another as can be identified.

1.12 <u>Projecting an Unbounded Line Segment onto N</u>

Let us <u>project</u> an <u>unbounded line segment</u> emanating from $\bar{z}$ in the

<u>reference direction</u> $d \in R^k$ onto N. Rearranging the augmented weighted

Tchebycheff sampling program and substituting $\bar{z} + \theta d$ for z**, where θ

goes 0 → +∞, we have the <u>augmented weighted Tchebycheff parametric</u>

<u>sampling program</u>

$$\min \{\alpha - \rho \sum_{i-1}^{k} z_i\}$$

$$\text{s.t.} \quad \alpha \geq \lambda_i(\bar{z}_i + \theta d - z_i) \qquad i \in K$$

$$f_i(x) = z_i \qquad i \in K$$

$$x \in S$$

$\alpha \in R$ unrestricted, $z \in R^k$ unrestricted

whose solution as θ goes $0 \rightarrow +\infty$ is the projection of the unbounded line segment onto N.

1.13 Contracting Λ-Space

In some of the interactive procedures, convergence to a final solution is controlled by contracting Λ-space as follows. Let $\lambda^{(h)} \in \Lambda$ be the Tchebycheff vertex λ-vector defined by the current solution $z^{(h)}$ and z** after iteration h. Then, for use on iteration h+1, we construct the interval defined subset of weighting vector space about $\lambda^{(h)}$

$$\Lambda^{(h+1)} = \{\lambda \in \Lambda | \lambda_i \in (\ell_i^{(h+1)}, \mu_i^{(h+1)}), \sum_{i=1}^{k} \lambda_i = 1\}$$

Using to a routine by Liou (1984), we can control the size of $\Lambda^{(h+1)}$. For instance, if we want $\Lambda^{(h+1)}$ to be, say, 36% of the size of $\Lambda^{(h)}$, Liou's routine will compute the half-width r so that we can construct the intervals

$$(\ell_i^{(h+1)}, \mu_i^{(h+1)}) = \begin{cases} (0, 2r^h) & \text{if } \lambda_i^{(h)} - r^h \leq 0 \\ (1 - 2r^h, 1) & \text{if } \lambda_i^{(h)} + r^h \geq 1 \\ (\lambda_i^{(h)} - r^h, \lambda_i^{(h)} + r^h) & \text{otherwise} \end{cases}$$

where r^h is r raised to the h-th power such that $\Lambda^{(h+1)}$ is 36% of the size of Λ.

PART 2: CURRENT STATUS

2.1 Interactive Procedures

The most prominent interactive procedures for interactive multiple objective programming are as follows:

Non-Reference Point Procedures

1.   e-Constraint Method (traditional)
2.   Geoffrion-Dyer-Feinberg Procedure (1972)
3.   Zionts-Wallenius Procedure (1976 and 1983)
4.   Interactive Weighted-Sums Method (see Steuer (1986, Section 13.5))
5.   Interactive Surrogate Worth Tradeoff Method (Chankong and Haimes (1978 and 1983))

Reference Point Procedures

6.   STEM (Benayoun, de Montgolfier, Tergny and Larichev (1971))
7.   Wierzbicki's Aspiration Criterion Vector Method (1977, 1982 and 1986)
8.   Interactive Goal Programming (see Franz and Lee (1980))
9.   Tchebycheff Method (Steuer and Choo (1983))
10.   Satisficing Tradeoff Method (Nakayama and Sawaragi (1984))
11.   VIG: Visual Interactive Approach (Korhonen and Laakso (1986)) and Pareto Race (Korhonen and Wallenius (1988))

2.2 Descriptions of Reference Point Procedures

STEM begins by constructing a payoff table and then uses z* as a reference criterion vector. Using its own rules to compute a λ-vector, the first stage only of the lexicographic sampling program is solved to

make a single probe of the nondominated set.   Determining which
components of the resulting criterion vector are to be relaxed so that
others can be improved, constraints are constructed to reduce the
feasible region and a new λ-vector is computed.   Using the new λ-vector,
the first stage of the lexicographic sampling problem is again solved to
produce a new criterion vector.   Determining which new components are to
be relaxed to afford improvement in the others, the feasible region is
again reduced and another λ-vector is computed, and so forth.

Wierzbicki's Aspiration Criterion Vector Method begins by
constructing a payoff table, establishing a z** reference criterion
vector, and having the DM specify an aspiration criterion vector $q^{(1)}$.
Then the augmented weighted Tchebycheff sampling program is solved using
the Tchebycheff vertex λ-vector defined by $q^{(1)}$ and z**.   After
examining the solution produced, the DM is asked to specify an updated
aspiration criterion vector $q^{(2)}$.   Using $q^{(2)}$ and z** to determine a new
Tchebycheff vertex λ-vector, the augmented weighted Tchebycheff program
is again solved.   After examining the solution produced, the DM is asked
to specify another updated aspiration criterion vector $q^{(3)}$, and so
forth.

In Interactive Goal Programming we set priority levels, penalty
weights, and target values.   Then, from the solution generated, it is
hoped that some revised problem configuration will suggest itself that
will lead to a better solution, and so forth.   This is an ad hoc method
because it is never precisely clear how to configure the problem and one
must proceed on instinct.

The Tchebycheff Method begins by establishing a z** reference
criterion vector.   Then a group of evenly dispersed λ-vectors from $\Lambda^{(1)}$

$= \Lambda$ is obtained and the lexicographic sampling program is solved for each of the λ-vectors. From the resulting criterion vectors, the DM selects his or her most preferred $z^{(1)}$. Using $z^{(1)}$ and z^{**}, a Tchebycheff vertex λ-vector is computed. About this λ-vector, a interval defined subset $\Lambda^{(2)}$ of weighting vector space is formed. Then a group of evenly dispersed λ-vectors is drawn from $\Lambda^{(2)}$, and for each of the λ-vectors, the lexicographic sampling problem is again solved. About the Tchebycheff vertex λ-vector pertaining to the most preferred of the generated criterion vectors $z^{(2)}$, a smaller interval defined subset $\Lambda^{(3)}$ of weighting vector space is formed, and so forth.

The <u>Satisficing Tradeoff Method</u> begins by establishing a z^{**} reference criterion vector and having the DM specify an aspiration criterion vector $q^{(1)} < z^{**}$. Then the augmented weighted Tchebycheff sampling program is solved using the Tchebycheff vertex λ-vector defined by $q^{(1)}$ and z^{**} to produce $z^{(1)}$. The DM then specifies which components of $z^{(1)}$ are to be increased, the amounts of each increase, and which components are to be relaxed. Using dual variable information available at $z^{(1)}$, the amounts of relaxation are determined to form a second aspiration criterion vector $q^{(2)}$. Using the new Tchebycheff vertex λ-vector defined by $q^{(2)}$ and z^{**}, the augmented weighted Tchebycheff sampling program is solved to produce $z^{(2)}$. The DM then specifies which components of $z^{(2)}$ are to be increased, the amounts of each increase, and so forth.

With reference to some starting criterion vector $z^{(0)}$, <u>VIG</u> begins by asking the DM to specify an aspiration criterion vector $q^{(1)}$. Then, using the augmented weighted Tchebycheff parametric sampling program, the unbounded line segment emanating from $z^{(0)}$ through $q^{(1)}$ is projected onto the nondominated set. Using computer graphics, the <u>trajectories</u> of criterion values describing the projection of the line segment onto N are displayed. After viewing the trajectories, the DM specifies his or her most preferred criterion vector along the projected line segment to yield $z^{(1)}$. At this point the DM specifies a new aspiration criterion vector $q^{(2)}$. Then, again using the augmented parametric sampling program, the unbounded line segment emanating from $z^{(1)}$ through $q^{(2)}$ is projected onto N. After viewing the trajectories pertaining to the new projected line segment, the DM selects the next criterion vector $z^{(2)}$. The DM now specifies a third aspiration criterion vector $q^{(3)}$, and so forth. The differences between VIG and <u>Pareto Race</u> are as follows:

(1)   An aspiration criterion vector is specified by the DM only at the beginning of the procedure in order to obtain a starting criterion vector.

(2)   An initial <u>reference direction</u> is obtained using objective function ranges as specified by the DM. After this, new reference directions are calculated in response to the DM's use of special function keys.

(3)   The unbounded line segment from the current criterion vector in the reference direction is not projected onto N. Only the projections of discrete steps along the reference direction are projected onto N each iteration.

2.3   Implementation Similarities

While the literature has stressed differences among the various

interactive procedures, most of the procedures, particularly of the
reference point variety, have remarkable implementation similarities.
For instance, they are all described by the following general
algorithmic outline:

Step 1:   Set the controlling parameters for the 1-st
            iteration.

Step 2:   Solve the lexicographic, augumented parametric,
            or augmented weighted Tchebycheff sampling program
            one or more times to probe the nondominated set.

Step 3:   Examine the criterion vector results.

Step 4:   If the DM wishes to continue, go to Step 5.
            Otherwise, exit with the current solution.

Step 5:   Set the controlling parameters for the next
            iteration and go to Step 2.

Controlling parameters are parameters in the sampling programs
that are varied in order to sample different points from the
nondominated set. Note that all steps, except Step 2, are free of heavy
number-crunching demands.

PART 3:   FUTURE DIRECTIONS

3.1   Consolidation and the Unified Sampling Program

Consolidation refers to the combining of different interactive
procedures into a common computer package. This is possible because (a)
all of the reference point interactive procedures of this paper fit the
general algorithmic outline and (b) all of the Step 2 sampling programs
are special cases of the unified sampling program

$$\text{lex min } \{s_1(\alpha,z,d^-,d^+),\ s_2(\alpha,z,d^-,d^+),\ldots,\ s_L(\alpha,z,d^-,d^+)\}$$

$$\text{s.t.} \qquad \alpha \geq \lambda_i(z_i^{**} + \theta d_i - z_i) \qquad i \in G$$

$$z_i \geq e_i \qquad\qquad i \in H$$

$$z_i + d_i^- \geq t_i \qquad\qquad i \in I$$

$$z_i - d_i^+ \leq u_i \qquad\qquad i \in J$$

$$f_i(x) = z_i \qquad\qquad i \in K$$

$$x \in S$$

$$d^-, d^+ \geq 0$$

$$\alpha \in R \text{ unrestricted, } z \in R^k \text{ unrestricted}$$

where:

$s_m(\alpha,z,d^-,d^+)$ <u>scalar-valued function</u> of m-th lexicographic level

L   number of lexicographic levels

z^{**} reference criterion vector (or current criterion vector in VIG and Pareto Race)

d   <u>reference direction</u>

t   <u>satisficing</u> target vector

u   <u>saturation</u> target vector

$K = \{1,\ldots,k\},\ G \subset K,\ H \subset K,\ I \subset K \text{ and } J \subset K$

The scalar-valued function of m-th lexicographic level is given by

$$s_m(\alpha,z,d^-,d^+) = \sigma_{(m)}\alpha - \rho_{(m)}\sum_{i=1}^{k} z_i + \tau_{(m)}\left[\sum_{i \in I} w_{(m)i}^- d_i^- + \sum_{i \in J} w_{(m)i}^+ d_i^+\right]$$

where $\sigma_{(m)}$, $\rho_{(m)}$ and $\tau_{(m)}$ are <u>procedure-dependent constants</u> which serve primarily as "switches" to include portions of the scalar-valued functions, and the $w_{(m)}^-$ and $w_{(m)}^+$ are vectors of penalty weights for deviations from target values. The controlling parameters which may be manipulated in the unified sampling program are λ, z^{**}, d, θ, e, t, u, $w_{(m)}^-$ and $w_{(m)}^+$, for m = 1,...,L.

3.2   Customizing the Unified Sampling Program

To illustrate how the unified sampling program can be customized for different procedures, consider STEM. By letting

    (a)   L = 1

    (b) $\sigma_{(1)} = 1$, $\rho_{(1)} = 0$ and $\tau_{(1)} = 0$

    (c)   G = K, $z^{**} = z^*$ and $\theta = 0$

    (d)   I = J = ϕ

    (e)   H $\subset$ K as determined by the rules of STEM.

we have the sampling program of STEM:

$$\min \{\alpha\}$$
$$\text{s.t. } \alpha \geq \lambda_i(z_i^* - z_i) \qquad i \in K$$
$$z_i \geq e_i \qquad i \in H$$
$$f_i(x) = z_i \qquad i \in K$$
$$x \in S$$
$$z \in R^k \text{ unrestricted}$$

With regard to the Tchebycheff Method, let

    (a)   L = 1

    (b) $\sigma_{(1)} = 1$, $\rho_{(1)} = \rho$ and $\tau_{(1)} = 0$

    (c)   G = K and $\theta = 0$

    (d)   H = I = J = ϕ

Thus we have the augmented weighted Tchebycheff sampling program

$$\min \{\alpha - \rho \sum_{i=1}^{k} z_i\}$$

$$\text{s.t. } \alpha \geq \lambda_i (z_i^{**} - z_i) \qquad i \in K$$

$$f_i(x) = z_i \qquad i \in K$$

$$x \in S$$

$\alpha \in R \text{ unrestricted, } z \in R^k \text{ unrestricted}$

The unified sampling program is customized similarly for the sampling programs of the other reference point procedures.

3.3 Workhorse Software

A common computer package for interactive multiple objective programming would essentially be a supervisory program for conducting Steps 1, 2, 3 and 5 of the general algorithmic outline. For Step 2 the supervisory program would call workhorse software as a subroutine to obtain solutions from the unified sampling program. In the case of a multiple objective linear program, workhorse software would be a conventional (single criterion) commercial-grade LP program product. In the case of integer and nonlinear multiple objective programs, we would use conventional integer programming and nonlinear programming codes.

3.4 Procedure-Switching

One of the results of the empirically-based research of Brockhoff (1985) and Buchanan and Daellenbach (1987) is that users knowledgeable of interactive multiple objective programming may wish to use different procedures on different iterations. We refer to this as

procedure-switching. For instance, a user may wish to start with the

Tchebycheff Procedure to sample the entire nondominated set, then switch

to Pareto Race to explore a portion of the nondominated set, and finally

switch to Wierzbicki's Aspiration Criterion Vector Method to pinpoint a

final solution. To accomodate procedure-switching, Steps 1 and 5 of the

general algorithmic outline are modified as follows:

> Step 1:   Set the controlling parameters of the selected
>   procedure for the 1-st iteration.
>
> Step 5:   Set the controlling parameters of the selected
>   procedure for the next iteration and go to Step 2.

3.5   Cognitive Equilibrium

The conventional decision-making paradigm is that the DM possesses

a utility function that he or she wishes to maximize over the feasible

region. Difficulties have been experienced with this paradigm in

interactive multiple objective programming. We know that the DM's

aspirations change in many multiple criteria problems as more is learned

about the problem. Does this mean that the DM's utility function can

make major changes over a short period of time and is thus unstable? Or

does this mean that a DM doesn't truely understand his or her utility

function without significant interaction with the problem implying, that

utility functions are much more difficult to assess a priori that

previously thought?

A new decision-making paradigm proposed by Zeleny (1989) called

cognitive equilibrium seems more appropriate to interactive multiple

objective programming. In this paradigm, decision making is the process of recursively redefining the problem and redefining one's aspirations until a form of stability or equilibrium is obtained at which time the problem is solved. In other words, decision making is the process of searching for harmony among chaos. It appears that this concept of decision making goes a lot further to explain what the field of interactive multiple objective programming has been encountering than conventional utility theory.

3.6   Network Optimization Applications

Consider the linear program (LP)

$$\max \ \{c^T x \,|\, x \in S\}$$
$$\text{s.t.} \quad S = \{x \in R^n \,|\, Ax = b, \ x \geq 0, \ b \in R^m\}$$

If, apart from upper and lower bounds on the variables, A has (i) no more than two nonzero elements per column and (ii) the nonzero elements are either 1's or -1's, the LP is a network. Because of the special structure of A, networks can be solved using high-speed solution procedures.

Let us now consider a multiple objective network. Unfortunately, with the interactive procedures discussed in this paper, the unified sampling program is not a network because of (non-network) side-constraints. However, with the development of high-speed codes that can handle limited numbers of side-constraints (Kennington and Whisman (1986)), multiple objective networks offer a growth area for applications in the 1990s.

3.7   Russian Interactive Research

There is a significant body of research on interactive multiple objective programming in the Russian literature about which little is known in the West. Fortunately, an English-written survey of this research appears in Lieberman (1990). It is expected that the separateness of research in the East and West will begin moving to full integration at a rapid pace.

REFERENCES

[1]   Benayoun, R., J. de Montgolfier, J. Tergny, and O. Larichev (1971). "Linear Programming with Multiple Objective Functions: Step Method (STEM)," Mathematical Programming, Vol. 1, No. 3, pp. 366-375.

[2]   Brockhoff, K. (1985). "Experimental Test of MCDM Algorithms in a Modular Approach," European Journal of Operational Research, Vol. 22, No. 2, pp. 159-166.

[3]   Buchanan, J. T. and H. G. Daellenbach (1987). "A Comparative Evaluation of Interactive Solution Methods for Multiple Objective Decision Models," European Journal of Operational Research, Vol. 29, No. 3, pp. 353-359.

[4]   Chankong, V. and Y. Y. Haimes (1978). "The Interactive Surrogate Worth Trade-off (ISWT) Method for Multiobjective Decision-Making," Lecture Notes in Economics and Mathematical Systems, Vol. 155, Springer-Verlag, pp. 42-67.

[5]   Chankong, V. and Y. Y. Haimes (1983). Multiobjective Decision Making: Theory and Methodology, New York: North-Holland.

[6]   Franz, L. S. and S. M. Lee (1980). "A Goal Programming Based Interactive Decision Support System," Lecture Notes in Economics and Mathematical Systems, Vol. 190, Springer-Verlag, pp. 110-115.

[7]   Gardiner, L. R. (1989). "Unified Interactive Multiple Objective Programming," Ph.D. Dissertation, Department of Management Science & Information Technology, University of Georgia, Athens, Georgia, USA.

[8]   Geoffrion, A. M., J. S. Dyer, and A. Feinberg (1972).   "An
      Interactive Approach for Multicriterion Optimization, with an
      Application to the Operation of an Academic Department,"
      Management Science, Vol. 19, No. 4, pp. 357-368.

[9]   Kennington, J. and A. W. Whisman (1986).   "NETSIDE Users Guide",
      Technical Report 86-OR-01, Department of Operations Research,
      Southern Methodist University, Dallas, Texas 75275.

[10] Korhonen, P. J. and J. Laakso (1986).   "A Visual Interactive
      Method for Solving the Multiple Criteria Problem," European
      Journal of Operational Research, Vol. 24, No. 2, pp. 277-287.

[11] Korhonen, P. J. and J. Wallenius (1988).   "A Pareto Race," Naval
      Research Logistics, Vol. 35, No. 6, pp. 615-623.

[12] Kreglewski, T., J. Paczynski, J. Granat, and A. P. Wierzbicki
      (1988).   "IAC-DIDAS-N: A Dynamic Interactive Decision Analysis
      and Support System for Multicriteria Analysis of Nonlinear Models
      with Nonlinear Model Generator Supporting Model Analysis,"
      WP-88-112, International Institute for Applied Systems Analysis,
      Laxenburg, Austria.

[13] Lewandowski, A. and M. Grauer (1982).   "The Reference Point
      Approach: Methods of Efficient Implementation," WP-82-26,
      International Institute for Applied Systems Analysis, Laxenburg,
      Austria.

[14] Lewandowski, A., T. Kreglewski, T. Rogowski, and A. P. Wierzbicki
      (1987).   "Decision Support Systems of DIDAS Family (Dynamic
      Interactive Decision Analysis and Support)," Archiwum Automatyki i
      Telemechaniki, Vol. 32, No. 4, pp. 221-246.

[15] Lieberman, E. R. (1990).   Multi-Objective Programming in the USSR,
      book manuscript, School of Management, State University of New
      York at Buffalo, Buffalo, New York, USA.

[16] Liou, F. H. (1984).   "A Routine for Generating Grid Point Defined
      Weighting Vectors," Masters Thesis, Department of Management
      Science & Information Technology, University of Georgia, Athens,
      Georgia, USA.

[17] Nakayama, H. and Y. Sawaragi (1984).   "Satisficing Trade-off
      Method for Multiobjective Programming." Lecture Notes in
      Economics and Mathematical Systems, Vol. 229, Springer-Verlag, pp.
      113-122.

[18] Rogowski, T., J. Sobczyk, and A. P. Wierzbicki (1988).
      "IAC-DIDAS-L: Dynamic Interactive Decision Analysis and Support
      System: Linear Version," WP-88-110, International Institute for
      Applied Systems Analysis, Laxenburg, Austria.

[19] Steuer, R. E. (1986). Multiple Criteria Optimization: Theory,
      Computation, and Application, (published by John Wiley & Sons, New
      York; republished by Krieger Publishing, Melbourne, Florida), 546
      pp.

[20] Steuer, R. E. (1990). "ADBASE Operating Mnaual," Department of
      Management Science & Information Technology, University of
      Georgia, Athens, Georgia, USA.

[21] Steuer, R. E. and E.-U. Choo (1983). "An Interactive Weighted
      Tchebycheff Procedure for Multiple Objective Programming,"
      Mathematical Programming, Vol. 26, No. 1, pp. 326-344.

[22] Wierzbicki, A. P. (1977). "Basic Properties of Scalarizing
      Functionals for Multiobjective Optimization," Mathematische
      Operationsforschung und Statistik - Series Optimization, Vol. 8,
      No. 1, pp. 55-60.

[23] Wierzbicki, A. P. (1982). "A Mathematical Basis for Satisficing
      Decision Making," Mathematical Modelling, Vol. 3, pp. 391-405.

[24] Wierzbicki, A. P. (1986). "On the Completeness and
      Constructiveness of Parametric Characterizations to Vector
      Optimization Problems," OR Spektrum, Vol. 8, No. 2, pp. 73-87.

[25] Zeleny, M. (1989). "Stable Patterns from Decision-Producing
      Networks: New Interfaces of DSS and MCDM," MCDM WorldScan, Vol.
      3, Nos. 2 & 3, pp. 6-7.

[26] Zionts, S. and J. Wallenius (1976). "An Interactive Programming
      Method for Solving the Multiple Criteria Problem," Management
      Science, Vol. 22, No. 6, pp. 652-663.

[27] Zionts, S. and J. Wallenius (1983). "An Interactive Multiple
      Objective Linear Programming Method for a Class of Underlying
      Nonlinear Utility Functions," Management Science, Vol. 29, No. 5,
      pp. 519-529.

ASPIRATION-LED DECISION SUPPORT SYSTEMS: THEORY AND METHODOLOGY

A.P. Wierzbicki

Warsaw University of Technology, Warsaw, Poland

ABSTRACT

This paper presents a review of the theory and methodology of aspiration-led multi-objective decision support systems as developed during the last decade. After a short historical note on the development of decision analysis and multi-objective optimization, diverse ways of understanding the concept of rationality in decision analysis and support are discussed and the development of aspiration-led decision support systems (ALDSS) is outlined. Foundations of multi-objective optimization theory are reviewed and their relations to reference point optimization, achievement scalarizing functions and aspiration-led decision support are presented, together with their extensions to the problems of dynamics, uncertainty, multi-person decisions. Various aspects of methodology of aspiration-led decision support are discussed along with the description of theoretical results. Some further aspects, such as the phases of decision support, the questions of possible standards of model computerization, the issues of optimization tools, of interactive graphics for decision support, are only outlined. Instead of conclusions, topics for further studies are indicated.

1. HISTORICAL DEVELOPMENT AND BASIC CONCEPTS.

1.1. Historical development of decision analysis and support.

During last fifty years there was a rapid development of decision analysis together with related multi-objective and dynamic optimization as well as game-theoretical aspects. All this development was connected with the discovery and explosive growth of computer technology. Various problems related to mathematical optimization, to rational economic behavior, to decisions resulting from negotiations, bargaining and compromise, were in fact studied much earlier. But the possibility of applying more sophisticated methods of decision evaluation and optimization supported by computer technology was the decisive stimulus for the development of modern decision analysis and support.

The research in this field was by no means restricted to applied mathematics and computer science; most development impulses came from related fields. Economic planning motivated L. Kantorovich and military logistics - G. Dantzig to develop optimization techniques known as linear programming[1]. This has led to many generalizations and to the development of a broad field called mathematical programming that uses optimization and model simulation techniques together with computer technology; the issues of multi-objective decision optimization were included already in the early works of H.W. Kuhn and A.W. Tucker. The development of mathematical programming was further stimulated by several important fields of applications including scientific research and development, engineering design and so called operations research related mostly to management in business and administration. Later it was also applied to environmental and various socio-economic problems.

Research questions arising in control engineering and, more generally, in systems theory, motivated first N. Wiener and W.R. Ashby, then A.A. Feldbaum and L.S. Pontryagin, R.E. Bellman and R. Kalman to develop mathematical control theory and dynamic systems theory as more advanced parts of mathematical programming; today, they are also essential elements of contemporary decision analysis and support.

The questions of economic rationality of decisions made by many individuals that are acting independently motivated J. von Neumann and O. Morgenstern, J. Nash and many others to develop game theory, originally as a tool for a better understanding of functioning of economic markets.

- - - - - - -

[1] We do not give here references to publications and authors that should be widely known from university courses on mathematical optimization and systems theory.

But the tools of game theory were soon used, by H. Raiffa and others, to develop basic elements of multi-objective decision analysis. Other disciplines, such as psychology or diplomatic negotiation theory, contributed also to a better understanding of decision rationality.

At the same time mathematical programming, operations research and systems theory contributed jointly to the development of the theory and various techniques of multi-objective optimization, through the researches of T.C. Koopmans, K.J. Arrow, L.A. Zadeh, A. Charnes and W.W. Cooper, N.O. da Cunha and E. Polak, A.M. Geoffrion, C. Olech, V.L. Volkovich, M.E. Salukvadze, O.I. Larichev, B. Roy, P.L. Yu, to mention only some of the early researchers in this field until the beginnings of 1970-ties; see [1], [2] for a review. Since that time - during the last 20 years - the theory, techniques and methodology of multi-objective optimization and decision making has been intensively advanced, culminating in many monographs such as by Y. Sawaragi, H. Nakayama and T. Tanino [1], P.L. Yu [2], R.E. Steuer [3], F. Seo and M. Sakawa [4].

A new trend that started after 1970 was the development of decision support systems (DSS), originated independently of decision analysis and optimization. It was understood initially as a next, more advanced phase of computerized information processing, stimulated by the rapid development of computer technology, software and applications, and especially by the emerging concept of their user-friendliness, see e.g. P.G.W. Keen and M.S. Scott Morton [5]. Soon, concepts of artificial intelligence based on logical modeling and inference rules were included in decision support systems, see e.g. [6]. However, it was also realized that a full development of such systems required the inclusion of the results on multi-objective decision analysis and optimization.

This particular trend was especially supported by the research done in the International Institute for Applied Systems Analysis (IIASA) in Laxenburg near Vienna and its network of cooperating institutions. Many conferences organized by IIASA in the field of multi-objective optimization, decision analysis and support, as well as some research done in IIASA - see e.g. [7], [8] - contributed to the perception that optimization and decision analysis should not be understood as the methods of substituting human decision makers by computerized, impersonal software packages, but much rather they should and can be used as the tools of developing more sophisticated but user-friendly support for humans facing novel decision situations or striving to master the complexity of interrelated decisions in modern world, while maintaining the sovereignty of their decisions. This perception is also related to a deeper understanding of the concept of decision rationality.

1.2. Frameworks of rationality.

The discussions between representatives of various cultural and professional backgrounds at IIASA have shown that the concept of rational decisions can have diverse interpretations and has cultural motivation - see e.g. the conference "Plural Rationality and Interactive Decision Processes" [9]. This understanding is particularly important when constructing decision support systems. A basic conclusion from this diversity and a principle to be followed is that a particular perception of rationality represented by the designer of a DSS should not be imposed on the user of this system (although many DSS designed until now do not quite follow this principle).

An important distinction is between the calculative or analytical rationality and the deliberative or holistic rationality, the hard versus soft approach. S.E. Dreyfus - see e.g. [10] - made a most consistent argument for the soft approach by showing that any individual way of decision making depends on the level of expertise attained through learning. A novice needs calculative rationality, while a master expert does not need it - at least, in a routine decision situation; he arrives at best decisions immediately by absorbing and intuitively processing all available information, presumably in a parallel processing scheme but in a way that is unknown until now (consider, for example , a driver in a formula I car race). When facing novel situations, the thinking of a master expert is characterized by searching for "new angles", culminating in the "aha" or heureka effect of perceiving a new perspective. The arguments for holistic, soft decision making originate thus from a culture of experts.

Dreyfus does not, however, follow his reasoning to its logical end: even a master expert might need calculative decision support, either in order to simulate and learn about a novel decision situation, or to fill in the details of an outline of his intuitive decision in a repetitive but complex situation. Novice decision makers need calculative decision support in order to learn and become experts; experts in a given field might be novices in other disciplines.

In the field of calculative rationality there are several approaches or frameworks that in fact also reflect the cultural background of perceiving what is rational. The utility maximization framework, characteristic for economic - theoretical understanding of rationality, has been long accepted as a basis of decision analysis and is often incorrectly interpreted as an absolute, universal basis for rational decisions. While its theoretical aspects are most developed and it describes well practical mass economic behavior, its power for predicting individual decisions is very limited. It might be understood as expressing the rationality and culture of small entrepreneurs facing an infinite market.

In the utility maximization framework it is, however, difficult to account for various levels of expertise and to support learning; a "perfectly rational" individual is supposed to be consistent, that is, not to make mistakes (which are essential for learning) and not to change his taste nor mind. This probably explains the failures of early variants of decision support systems that attempted to approximate the utility function of a user and then to suggest a decision that maximizes such approximated utility function. It takes many questions and answers to approximate a utility function - and, when the user finally learns something new from the decision support system, his utility might change and the lengthy process must be repeated. Some users resent too detailed questions about their utility or simply refuse to think in terms of utility maximization. In such a situation, trying to follow the economic-theoretical principles of normative, predictive and descriptive decision analysis and to convince the user that he "should" maximize his utility and be consistent results actually in violating user's sovereignty as a decision maker.

The normative theory of utility maximization has been criticized by the behavioral or descriptive school that postulated the satisficing rationality framework - see e.g. H. Simon [11], [12]. An important contribution of this school is the recognition that individual decision makers develop adaptively aspiration levels and use these aspirations to guide their decisions. Very often, they cease to maximize when reaching decision outcomes consistent with their aspirations; thus, they make satisficing decisions. Such model of decision behavior has been long since confirmed by many experimental studies - see e.g. [13]. However, the strength of the normative thinking in economics is best illustrated by the fact that even now a satisficing behavior is called "bounded rationality" in experimental economic studies - implying that it is not a perfect but only a practical concept.

On the other hand, some more radical thinkers have perceived - see e.g. J.K. Galbraith [14] - that the satisficing rationality is not a bounded, but a different rationality concept, typical for the culture of big industrial and administrative organizations where the decision makers face complex multi-person situations. Originally, three reasons for abandoning maximization and using satisficing decisions instead were quoted: the difficulty and the cost of computing optimal decisions; the necessity to account for uncertainty effects; and the complexity of multi-personal relations. Moreover, the behavioral school was critical about the use of mathematical programming, since this use was interpreted as equivalent to utility maximization. Today, we see that mathematical programming can be as well used to help in computing satisficing decisions, and the two first reasons for not maximizing turned out to be immaterial: the cost of optimization when using modern computers is negligible and experts can anyway optimize intuitively better than computers; similarly, we have developed many ways of dealing with uncertainty. But the research on so called rationality traps and on the evolution of cooperation - see A. Rappoport [15], R. Axelrod [16] - has confirmed the importance of abandoning short-term utility maximization in complex multi-person decision situations.

However, when trying to incorporate the results of such research into decision support systems, an important question is: in which case should a decision maker stop to maximize upon reaching aspiration levels? He should do it for good specific reasons, such as avoiding rationality traps or conflict escalation; but the decision to stop maximization should be his own, not imposed on him by a DSS reflecting the rationality of its designer. Thus, for example, multi-objective decision support based on goal programming techniques - see e.g. [17], [18] - correctly represents satisficing rationality but fails to support other possible frameworks of rational behavior.

Beside utility maximization and satisficing, there are other frameworks. For example, the program- and goal-oriented planning and management framework - see V.M. Glushkov [19], G.S. Pospelov and V.A. Irikov [20] - represents the culture of planning (by which we should not understand only central economic command planning, see also [21]).

Utility maximization, or satisficing, or program- and goal-oriented planning are perceptions of rationality that originated in European or North-American culture. Though these perceptions might have some universal features, it is quite probable that other cultures might be also represented by different perceptions of rationality - say, the Japanese and Chinese culture might be represented by the rationality of achieving the best harmony (wa). The conclusion is that we should avoid even implicit cultural imperialism - the remains of XIX-th century in our subconscious attitudes - and refrain from imposing our own perception of rationality on others, also when designing decision support systems.

1.3. Aspiration-led decision support systems.

The development of aspiration-led decision support systems (ALDSS) originally started with attempts to find user-friendly means of interaction with multi-objective optimization programs by using reference point optimization, and to extend the concept of satisficing decision making - see [7], [8]. However, these concepts soon turned out to be broader - thus, the term quasi-satisficing was used, see [22] - and resulted in an idea to construct DSS that are based on multi-objective optimization but can accommodate plural, various perceptions of rationality while using such means of interaction (aspiration levels) that are universal and easy to understand by the users of such systems.

The basic assumptions of an aspiration-led DSS are:

- an ALDSS should contain a so called substantive model of the decision situation that represents the best available expert knowledge on possible outcomes of various alternative decisions, usually in the form of an analytical model; an ALDSS might contain also a judgmental model of the decision situation that specifies in some sense the preferences of the user;

- the user of an ALDSS defines a decision problem related to a given substantive model by selecting the decision outcomes that correspond to his objectives and telling what to do with these objective outcomes - whether to maximize, or minimize, or stabilize them (the last possibility means maximizing below and minimizing above a given reference level, which actually corresponds to satisficing behavior); the definition of the problem might be considered a part of the judgmental model;

- the user interacts with an ALDSS by stating or modifying his aspiration levels (or reservation levels, or - generally - reference points) for each objective outcome, while the ALDSS behaves like a perfect supporting staff with the user as its boss - that is, it uses all available information and processing capabilities, including sophisticated optimization and graphic display techniques, to find decisions that are in a sense best attuned to the requirements of the boss, and to display the results in a user-friendly way.

The concept of aspiration levels (that would be good to achieve) or reservation levels (that are important to achieve) for decision outcomes is rather universal and easily understood by users - although it differs from the concept of constraining levels (that must be achieved) typical for optimization techniques. This should be stressed particularly in respect to reservation levels: reaching such levels is interpreted as a necessary condition for a positive decision, hence they are often (through a mental shortcut) represented by constraints. However, a thorough analysis of a decision situation should also account for the possibility that reservation levels are not attainable - and a perfect supporting staff should not report "your demands are inconsistent -

change them" to its boss, but much rather should find a decision that
best approximates the unattainable reservation levels and report "sorry,
the best that we can reach is...". On the other hand, if the reservation
levels are attainable for objectives that are not specified as
stabilized, a perfect supporting staff should not be satisfied with
decisions that just result in attaining these levels - the staff should
also check how far one can improve the reservation levels by admissible
decisions.

Thus, the aspiration and reservation levels or reference points
correspond to soft constraints represented by penalty terms in
optimization techniques, not to hard constraints. The user of an ALDSS
should also have the possibility of specifying hard constraints on some
of his objectives; but the principal interaction variables correspond to
soft constraints. This distinguishes an ALDSS from diverse variants of
methods of (hard) constraint perturbation in multi-objective optimization
- such as the surrogate worth trade-off method, see Y.Y. Haimes et al.
[23], or other, earlier methods of variable constraints, see O.I.
Larichev [24] or G. Fandel [25].

Although the concept of aspiration levels is typical for satisficing
rationality framework, an ALDSS represents satisficing rationality only
if the user specifies all objectives as stabilized - in which case the
system behaves similarly to a goal programming software package. Other
frameworks of rationality can be also supported: for example, by
specifying hard constraints on the most important of his objectives, the
user of an ALDSS can express his thinking in terms of goal- and
program-oriented planning.

But an ALDSS can also support utility maximization - either
indirectly or directly. An indirect support occurs if we assume that the
user behaves as if maximizing his utility - but he is an expert who can
intuitively perform this maximization himself, given all pertinent
information. Usually, however, he has only an approximate mental model of
the available decision alternatives, their constraints and consequences -
while this information is represented in more detail by the substantive
model included in an ALDSS. By maximizing intuitively his utility on the
mental model, the user can specify the desirable consequences of his
decision - in other words, the aspiration levels or reference point that
is then communicated to the ALDSS, together with the information which
decision outcomes are improved when maximized, which should be minimized
etc. The function of the ALDSS is then not to outguess the user about his
utility function, but to support best the user by checking which
decisions come close to or give even better results than required by the

reference point. In such interaction, the user learns more about the decision situation. Therefore, his utility and the corresponding reference point might change, but this does not matter - since he is able to maximize intuitively his unstated utility function, the main purpose of an ALDSS is to support his learning.

A direct support of utility maximization by an ALDSS is needed if the user is a novice or, at least, not an expert yet. An additional algorithm is then required that suggests to the user how to change his aspiration levels or reference points in order to converge to a maximum of his utility function. The known methods of multi-attribute utility assessment - see [26] - can be applied for this purpose. On the other hand, various modern optimization algorithms can be adapted to deal with even unstated nor directly estimated, possibly changing utility functions. But in order not to offend the user, such an algorithm should be supplied on specific request, not imposed on him.

The basic assumptions about the functions of an ALDSS stated above are results of experience gathered when actually constructing and implementing such systems. The first, experimental version of such a system was constructed by M. Kallio, A. Lewandowski and W. Orchard-Hays at IIASA [27] and applied to a problem of studying long-term relations of forestry and industry in Finland. Many other versions, called DIDAS systems (dynamic interactive decision analysis and support, stressing their applications to problems described with dynamic models), were developed by A. Lewandowski and M. Grauer at IIASA, also by T. Kreglewski and T. Rogowski, later J. Sobczyk, J. Paczynski and J. Granat at the Institute of Automatic Control, Warsaw University of Technology, with diverse applications both in IIASA and in Poland, see [28]. Other versions of ALDSS were developed in IIASA by M. Grauer, A. Lewandowski and L. Schrattenholzer [29], M. Grauer and E. Zalai [30], S. Messner [31], M. Strubegger [32], A. Lewandowski, S. Johnson and A. Wierzbicki [33] and others. In a cooperative research with IIASA, several Polish research institutions developed many versions of ALDSS - for discrete alternatives, for decision models with nonlinear constraints and nondifferentiable optimization, with various applications in research, engineering design or industrial planning - see [28].

However, the principles of aspiration-led DSS were developed also by other groups of researchers that adopted aspiration or reference levels as principal interaction variables. H. Nakayama and Y. Sawaragi [34] concentrated on methods of supporting modification of aspiration levels by trade-off analysis, see also [35]. R. Steuer and E.V. Choo [36] modified a weighted Chebyshev procedure to use aspiration levels as interaction variables. P. Korhonen and J. Laakso [37], see also P. Korhonen and J. Wallenius [38], used variable reference points to organize a directional search and visual interaction along Pareto surface. F. Seo and M. Sakawa - see e.g. [4], but also in their earlier research - used reference points or aspiration levels in interactive fuzzy approaches to multi-objective programming. All these methodological and partly theoretical developments were accompanied by testing prototype decision support systems on a broad variety of applications ranging from engineering design - of bridges, photographic lenses, electronic circuits - to various environmental, regional, micro- or macro-economic problems - see also [39].

Beside the practical advantages of interaction based on aspiration or reservation levels - variables that are easy to understand and interpret - there are also essential theoretical advantages of aspiration-led decision support systems. These advantages relate to some specific concepts in multi-objective optimization theory, such as the separation of the set of attainable outcomes and the positive cone, or the local controllability of a parametric characterization of efficient solutions.

2. MULTI-OBJECTIVE OPTIMIZATION THEORY AND RELATED METHODOLOGICAL ISSUES.

2.1. Basic formulations.

A mathematical model of a multi-objective optimization problem is usually defined in a form that is too restrictive for direct applications in a DSS - see [28] for more detailed discussion of this issue. While keeping this in mind, we shall start with this restrictive model form.

A set X of admissible decisions in a decision space $\mathcal{X}$ is assumed to be given. The decision space can be, generally, any topological space; we assume only that the set of admissible decisions is compact. Therefore, the concept of an admissible decision is rather broad: a decision can belong to a discrete set obtained by just listing possible decision alternatives, or to a set of continuum power in a finite dimensional space of possible values of decision variables, as it is typical in engineering design, or to an infinite-dimensional set, when the decision is interpreted as a profile or time-trajectory of repetitive decisions, or even a function or probability distribution (in an infinite-dimensional decision space we might need special techniques to compactify the set X).

The set X of admissible decisions can be also defined implicitly by specifying decision constraints. For this purpose, a space of constraints

$\mathcal{Z}$ with a partial (pre-)ordering relation $\leqslant$ and a constraining mapping $g:\mathcal{X} \rightarrow \mathcal{Z}$ are defined, while the admissible decisions $x \in X$ are such that $g(x) \leqslant z$, where $z \in \mathcal{Z}$ is a given element of the space of constraints. When formulating a substantive model of a decision situation, however, it is useful not to fix constraints a priori but to treat them as potential decision outcomes that might be later defined either as constraints or as objectives depending on the problem specification by the user.

An essential part of a model of multi-objective decision situation is the specification of decision outcomes that belong to an outcome space $\mathcal{Y}$, in which a partial ordering relation or, simply speaking, a sense of optimization is defined. The outcome space $\mathcal{Y}$ can be also of rather general nature. For applications in ALDSS we assume that it is a normed space (which assumption is, as discussed later, rather strong but useful). Although infinite-dimensional normed spaces can be also usefully considered, we restrict here the attention to the simplest case when $\mathcal{Y} = \mathbb{R}^p$, where p is the number of optimized objectives.

We assume also that the sense of optimization in the outcome space is defined by a positive cone C in this space. A standard form of the positive cone is:

$$C = \{y \in \mathbb{R}^p: y_i \geq 0, i = 1,\ldots p\} = \mathbb{R}^p_+ \qquad (1a)$$

and corresponds to the assumption that all outcomes or objectives are maximized (minimized objectives can be taken into account by changing their signs). Another example of positive cone:

$$C' = \{y \in \mathbb{R}^p: y_i \geq 0, i = 1,\ldots p'; y_i = 0, i = p'+ 1,\ldots p\} \qquad (1b)$$

represents less standard assumption that the first p' objective outcomes are maximized while the remaining ones are stabilized around given reference level. Both C and C' are pointed cones, the only subspace they contain is the trivial one $\{0\}$ - which is usually required from a positive cone. They are also closed convex cones, which is an important but not absolutely essential property of a positive cone; finally, C has nonempty interior while int C' = $\emptyset$ (an empty set). In more complicated, infinite-dimensional outcome spaces, it is usually also possible to represent the sense of optimization by appropriate pointed positive cones.

The sense of optimization in the outcome space is actually a part of judgmental model; an essential part of a substantive model of a decision situation is the specification of consequences of each admissible decision x   X. Thus, we assume that an outcome mapping $f: X \rightarrow \mathcal{Y}$ is given. This mapping should be also interpreted rather broadly. Often, the outcomes of admissible decisions cannot be fully determined and various models of uncertainty can be included in this mapping. If the uncertainty is represented by a probabilistic model, then the outcome mapping might include a specified sense of averaging or forming expectations. If the uncertainty is represented by other types of models, such as fuzzy sets or set-valued mappings, then the space of outcomes and the positive cone might be suitably re-defined. A part of outcomes might be assessed only by judgment, thus certain components of outcome mapping might be defined by experts during the decision process. The outcome mapping can consist of a complicated dynamic model with a suitably defined space of outcomes - and a simulation of an outcome might be a complex task in itself.

While keeping in mind its possible broad interpretation, we shall restrict our attention to the set Y of attainable outcomes, defined as:

$$Y = f(X) \subset \mathcal{Y} \tag{2}$$

It is usually assumed that the outcome mapping is continuous and thus the set Y is compact, although weaker assumptions are sufficient for the existence of multi-objectively optimal or efficient solutions and stronger assumptions - such as differentiability - are useful when actually computing such solutions.

Given a positive cone C, we define the set of efficient outcomes (called Pareto-optimal if $C = \mathbb{R}^p_+$) in a standard way:

$$\hat{Y} = \{y \in Y: Y \cap (\hat{y} + \tilde{C}) = \emptyset\}; \quad \tilde{C} = C\backslash\{0\} \tag{3a}$$

where $\hat{y} + \tilde{C}$ denotes the cone $\tilde{C}$ - that corresponds to a strict inequality in the outcome space - shifted by the vector $\hat{y}$. A strong inequality in the outcome space corresponds to the interior of the cone C (at least, if $C = \mathbb{R}^p_+$); thus, weakly efficient (or weakly Pareto-optimal) outcomes are defined as:

$$\hat{Y}^w = \{y \in Y: Y \cap (\hat{y} + \text{int } C) = \emptyset\} \tag{3b}$$

The concept of weakly efficient outcomes is typically too weak for applications (note that $\hat{Y}^w = Y$ if we use the cone C'). Most useful for applications is the concept of properly efficient outcomes that can be defined in various ways - see Kuhn and Tucker [40], Geoffrion [41], Henig

[42], Sawaragi et al. [1], also this author [43], [44], [45]. We shall use here one of the latter definitions of properly efficient outcomes with a prior bound ε (such that the corresponding trade-off coefficients have a bound that is a priori known, approximately $1/\varepsilon$, as opposed to a bound that exists but we do not know how large it is). For this purpose, we define first the closure of an ε-conical neighborhood of the cone C:

$$C(\varepsilon) = \{y \in \mathbb{R}^p: \text{dist}(y,C) \leq \varepsilon ||y||\} \tag{4a}$$

where any norm in $\mathbb{R}^p$ can be used and the distance of y from C is defined as a Haussdorf distance that uses a topologically equivalent norm (hence, in $\mathbb{R}^p$, any other norm can be used); we shall show later an example of a useful form of $C(\varepsilon)$. Properly efficient outcomes with prior bound ε can be defined as:

$$\hat{Y}^{p\varepsilon} = \{\hat{y} \in Y: Y \cap (\hat{y} + \text{int } C(\varepsilon)) = \emptyset\} \tag{4b}$$

This definition does not have the deficiencies of the definition (3b), because the cone $C(\varepsilon)$ has a nonempty interior even if int C is empty, as in the case of C'. The set of properly efficient outcomes in their more traditional sense (with an existential, not known a priori bound) can be defined as:

$$\hat{Y}^p = \bigcup_{\varepsilon > 0} \hat{Y}^{p\varepsilon}$$

Note that the three types of efficiency (3a), (3b), (4b) are defined rather abstractly, except for the specific examples of positive cones; therefore, these definitions can be used as well in infinite-dimensional outcome spaces.

There are several types of questions in multi-objective optimization. One is to analyze entire sets $\hat{Y}$, $\hat{Y}^w$, or $\hat{Y}^{p\varepsilon}$; but we can usually analyze only their general, qualitative properties. Another is to select one or several representative elements $\hat{y} \in \hat{Y}$ ($\hat{Y}^w$ or $\hat{Y}^{p\varepsilon}$) for a human decision maker to choose from, which is typical for ALDSS but also for other DSS using multi-objective optimization. Historically, many researchers tried also to find a universal, "objective" way of choosing an element $\hat{y} \in \hat{Y}$ that would make unnecessary the choice made by a human decision maker. However, it was then realized that such a decision automation is good for machines only - in repetitive, standard decision problems; moreover, there is no universal way of selecting efficient outcomes and decisions independently of particular context.

2.2. Parametric scalarization.

In all above types of problems, a general method of analyzing efficient outcomes is to introduce a parametric scalarizing function. A set $W \subset \mathbb{R}^p$ of controlling parameters $w \in W$ is defined; examples of such controlling parameters are weighting coefficients (actually the elements of a dual space to the space of outcomes) or reference points (composed of aspiration or reservation levels, the elements of the space of outcomes). A parametric scalarizing function is a continuous function $s: \mathbb{R}^p \times W \rightarrow \mathbb{R}^1$ that possesses the following sufficiency property: there is a nonempty set $W^s \subset W$ such that:

$$\hat{\Psi}(w) = \operatorname*{Arg\,max}_{y \in Y} s(y,w) \subset \hat{Y} \text{ for all } w \in W^s \tag{5a}$$

where Arg max denotes the set of maximal points of the function. We might substitute $\hat{Y}$ in (5a) by $\hat{Y}^w$ or $\hat{Y}^{p\varepsilon}$ (and W^s by W^{sw} or $W^{sp\varepsilon}$), thus obtaining a slightly weaker or stronger sufficiency property.

Some ways of introducing parametric scalarization in multi-objective optimization are more complicated, requiring additional constraints or p times repeated maximization in (5a) - see later comments. But without such modifications, a general way of obtaining the sufficiency property is to choose a scalarizing function that is monotone in an appropriate sense. If W^s denotes the set of such w that $y'' - y' \in \tilde{C} = C \setminus \{0\}$ implies $s(y'',w) > s(y',w)$ (such w that $s(.,w)$ is strongly monotone) then (5a) holds. If W^{sw} denotes the set of such w that $y'' - y' \in \operatorname{int} C$ implies $s(y'',w) > s(y',w)$ (such w that $s(.,w)$ is strictly monotone) then (5a) holds for $\hat{Y}^w$, W^{sw}. If $W^{sp\varepsilon}$ denotes the set of such w that $y'' - y' \in \operatorname{int} C(\varepsilon)$ implies $s(y'',w) > s(y',w)$ (such w that $s(.,w)$ is $C(\varepsilon)$-strictly monotone, which implies strong monotonicity) then (5a) holds for $\hat{Y}^{p\varepsilon}$, $W^{sp\varepsilon}$ - see [44], [45].

A parametric scalarizing function has also necessity property that might be either complete or incomplete. An incomplete necessity property means that for a given set $W^n \subset W$ there exists a nonempty subset $\tilde{Y} \subset \hat{Y}$ such that, if $\hat{y} \in \tilde{Y}$, then there exists $\hat{w} \in W^n$ for which:

$$\hat{y} \in \hat{\Psi}(\hat{w}) = \operatorname*{Arg\,max}_{y \in Y} s(y,\hat{w}) \tag{5b}$$

Again, we can substitute $\hat{Y}$ by $\hat{Y}^w$ or $\hat{Y}^{p\varepsilon}$ (and $\tilde{Y}$ by $\tilde{Y}^w$ or $\tilde{Y}^{p\varepsilon}$, W^n by W^{nw} or $W^{np\varepsilon}$) to obtain incomplete necessity property for weakly efficient or properly efficient outcomes. Since a scalarizing function has the sufficiency property, we can always take $W^n = W^s$ (similarly for W^{nw} or $W^{np\varepsilon}$) and obtain a nonempty $\tilde{Y}$ ($\tilde{Y}^w$ or $\tilde{Y}^{p\varepsilon}$) in the above definition. However, we might also choose larger sets W^n (W^{nw} or $W^{np\varepsilon}$) in order to get larger sets $\tilde{Y}$ ($\tilde{Y}^w$ or $\tilde{Y}^{p\varepsilon}$) that desirably should cover $\hat{Y}$ ($\hat{Y}^w$ or $\hat{Y}^{p\varepsilon}$), in which case we say that the necessity property is complete. If it is complete and $W^{nw} = W^{sw}$, or $W^{np\varepsilon} = W^{sp\varepsilon}$, or - equivalently - if:

$$\bigcup_{w \in W^{sp\varepsilon}} \hat{\Psi}(w) = \hat{Y}^{p\varepsilon} \tag{5c}$$

(similarly for W^{sw} and $\hat{Y}^w$), then we say that the scalarizing function $s(y,w)$, used when defining $\hat{\Psi}(w)$, completely characterizes parametrically the set of properly efficient outcomes with prior bound ε (or the set of weakly efficient outcomes).

If (5c) holds, it is indeed a characterization: the maximization of the scalarizing function is not only sufficient, but also necessary condition, since for every $\hat{y} \in \hat{Y}^{p\varepsilon}$ there exists a parameter vector $\hat{w} \in W^{sp\varepsilon}$ such that $\hat{y}$ maximizes $s(y,w)$ over $y \in Y$. We did not define a complete characterization for efficient outcomes, because it is known - see [1], or an impossibility theorem in [44] - that the set of efficient outcomes Y, without its prior knowledge nor repeated maximizations, can be only almost completely characterized - in such a way that:

$$\hat{Y} \subset \text{closure} \bigcup_{w \in W^s} \hat{\Psi}(w) \tag{5d}$$

Beside completeness, a characterization of efficient solutions by maxima of a scalarizing function can have various other properties - see [44]. Such a characterization is locally parametric controllable if the point-to-set mapping $\hat{\Psi}(w)$ is Lipschitz-continuous (in an appropriate Haussdorf distance sense); this is an especially important property which - together with completeness - means that any (properly or weakly) efficient outcome can be obtained by fine-tuning the parameter vector. We might use also more descriptive properties of parametric characterizations - their easy computability, interpretability of their parameters w, etc.

2.3. Basic classes of scalarizing functions.

There are several important classes of scalarizing functions. If $C = \mathbb{R}^p_+$, the most elementary is the weighted sum of objective outcomes:

$$s(y,w) = \lambda^T y = \sum_{i=1}^{p} \lambda_i y_i; \quad w = \lambda \tag{6a}$$

with:

$$w^s = int\ \mathbb{R}^p_+ = w^{sp} = w^{np}; \quad w^n = \mathbb{R}^p_+ \backslash \{0\} = w^{sw} = w^{nw} \tag{6b}$$

where we often normalize the values of weighting coefficients (under an implicit assumption that values of objective outcomes are also normalized, see later discussion):

$$\bar{\lambda}_i = \lambda_i / \sum_{i=1}^{p} \lambda_i \tag{6c}$$

If the attainable outcome set Y is convex, then the weighted sum completely characterizes weakly Pareto-optimal outcomes (with weighting coefficients that are nonnegative and not all equal zero), the properly Pareto-optimal outcomes (with positive weighting coefficients) and the properly Pareto-optimal outcomes with a prior bound ε (an additional restriction that $\bar{\lambda}_i \geq \varepsilon/(1+p\varepsilon)$ results in trade-off coefficients bounded by $1 + 1/\varepsilon$, see [45]). The Pareto-optimal outcomes are only almost completely characterized, with $w^n \neq w^s$ but $w^n \subset$ closure w^s. The main drawbacks of the weighted sum as a scalarizing function are, however, that the necessary conditions hold only under convexity assumptions (since they rely on arguments of separating convex sets by linear functions; the sufficient conditions rely on monotonicity and are thus independent on convexity) and, which is more important, that the resulting characterization is not locally parametric controllable even in the simplest case when the set Y is a convex polyhedron.

Another basic class of scalarizing functions are norms of distance of an attainable outcome point $y \in Y$ from a shifted utopia or ideal point, defined as any point $\bar{y} \in \bar{Y} \subset \mathbb{R}^p$ that dominates entire Y:

$$\bar{Y} = \{\bar{y} \in \mathbb{R}^p: Y \subset \bar{y} - C\} \tag{7a}$$

If $C = \mathbb{R}^p_+$, then $\bar{Y} = \bar{\bar{y}} + C$, where $\bar{\bar{y}}$ is so called utopia or ideal point obtained by maximizing subsequently all objectives and combining their maxima into one vector.

In order to use a norm in the outcome space it is necessary that various outcomes are comparable and can be summed. This would be a very strong restriction since typically various outcomes have quite different meaning and physical units. Thus, when using a norm we implicitly assume that the outcomes are normalized: for each outcome y_i there is an interval of bounds $[y_{i,low}; y_{i,upp}]$ given and the outcome is transformed to a dimensionless normalized outcome $y_{i,n}$:

$$y_{i,n} = y_i / (y_{i,upp} - y_{i,low}) \tag{7b}$$

But there is no standard, "objective" way to define intervals of bounds. Even if we accept the utopia point $\bar{\bar{y}}$ as a "natural" upper bound (if all outcomes are to be maximized) - which is also open to discussion - a "natural" definition of a lower bound is more difficult. If we just take the tight lower bound on the set Y (or, equivalently, assume that all outcomes are positive, as e.g. in [46]) then we include also inefficient outcomes, while only efficient ones are of interest. We should use rather the so called nadir point that is supposed to represent the tight lower bound on efficient outcomes - the maximal element between such $\tilde{y}$ that $Y \subset \hat{y} + C$. However, computing a tight lower bound on $\hat{Y}$ is a difficult computational problem and simplistic approaches to it give inaccurate results. Usually, the best what we can do is to rely first on estimates of outcome ranges supplied by the user or an expert and then to compute approximate utopia and nadir points.

The arbitrariness of the normalization (7b) is in fact equivalent to choosing some weighting coefficients for all outcomes - although it makes clearly more sense to use normalized weighting coefficients (6c) first after normalizing all outcomes to a common range. Thus, we can define:

$$s(y,w) = ||\bar{y}_n - y_n||; \quad y_{i,n} = \bar{\lambda}_i y_i / (y_{i,upp} - y_{i,low}) \tag{7c}$$

with $w = \lambda \in W^s = $ int $\mathbb{R}_+^p$. If we take equal $\bar{\lambda}_i = 1/p$, $\bar{y}_n = \bar{\bar{y}}_n$ and minimize the norm (7c) (we assumed the scalarizing functions to be maximized, but we can always change the sign in their definition), then we obtain an efficient outcome that is in some sense neutral and provides a good starting point for the interaction with the user of a DSS. But we should not attach any general significance to the neutral outcome - although we can, for example, investigate its relation to various concepts of cooperative solutions in game theory - because the sense of neutrality and the properties of this outcome depend on the norm used in defining (7c).

If we take any norm l_p with $1 \le p < \infty$, by minimizing (7c) we obtain an almost complete characterization of Pareto-optimal outcomes, provided $\bar{y} \in \bar{Y}$. If we assume $\bar{y} \in \text{int } \bar{Y}$ and take the Chebyshev norm l_∞, we obtain a complete characterization of weakly Pareto-optimal outcomes; with the same $\bar{y}$, if we take an augmented Chebyshev norm (the sum of the norm l_∞ and the norm l_1 multiplied by ε), we obtain a complete characterization of properly Pareto-optimal outcomes with prior bound ε. These characterizations do not depend on convexity assumptions.

The scalarizing functions constructed with norms of the distance from an ideal or utopia outcome were first introduced by the researchers in the USSR - see V.L. Volkovich [47] - for static multi-objective optimization and then extended to applications for dynamic multi-objective optimization of the final state, see M.L. Salukvadze [46]. The results of Salukvadze were then popularized and extended in the USA by P.L. Yu and G. Leitmann [49], by M. Zeleny [50], [51] and others as so called displaced ideal or compromise programming methods - but adapted again for the static case, while the originators of this idea remained unknown in the West.

Without requiring that $\bar{y} \in \bar{Y}$, the minimization of a norm (7c) has been used extensively in goal programming approaches - see e.g. [17], [18]. Goal programming actually uses both weighting coefficients and the vectors $\bar{y}$ - interpreted as goals - as parameters in the scalarizing function. However, if we do not assume $\bar{y} \in \bar{Y}$, it is rather difficult to obtain efficient solutions when minimizing (7c) - we need then convexity assumptions and repeated optimizations to check efficiency, or we must switch from minimizing to maximizing (7c) with additional constraints if $\bar{y} \in Y$, see e.g. [44].

A scalarizing method that relies exclusively on parameters $\bar{y}$ - interpreted usually as reservation levels, although they are in fact treated as hard constraining values in this case - is the method of constraint perturbations. We define the j-th perturbation function of Pareto-optimal outcomes as:

$$\tilde{h}_j(\bar{y}) = \max_{y \in \tilde{Y}^{(j)}(\bar{y})} y_j \tag{8a}$$

where:

$$\tilde{Y}^{(j)}(\bar{y}) = \{y \in Y: y_i \ge \bar{y}_i \text{ for all } i \ne j\} \tag{8b}$$

It is known - see [41], [52], [23], [53] - that an outcome $\hat{y} \in Y$ is Pareto-optimal, $\hat{y} \in \hat{Y}$, if and only if with $\bar{y} = \hat{y}$ we obtain $\tilde{h}_j(\bar{y}) = \hat{y}_j$ for all $j = 1,\ldots p$ when solving problems (8a,b). If Y is convex, an outcome $\hat{y} \in Y$ is properly Pareto-optimal (with existential bound), $\hat{y} \in \hat{Y}^p$, if and only if additionally the problems (8a,b) are <u>stable</u> for all $j = 1,\ldots p$ - that is, the functions $\tilde{h}_j(\bar{y})$ are Lipschitz-continuous with respect to $\bar{y}$ or, equivalently, the subdifferentials of these functions or the sets of Lagrange multipliers for the inequalities in (8b) are bounded.

This characterization, though it is complete for Pareto-optimal outcomes without convexity assumptions, is not quite satisfactory. If we do not know whether $\bar{y} \in Y$, the sets (8b) might become empty, which causes problems in computations. This characterization requires p repetitions of maximization; if we maximized only one of functions (8a), we could be sure to obtain that way only weakly Pareto-optimal outcomes. Moreover, problems (8a,b) are unstable for such Pareto-optimal outcomes that are not properly Pareto-optimal; this means that the corresponding computational problems are badly conditioned, it is not really practical to check this way Pareto-optimality of such outcomes.

Thus, the method of constraint perturbations is not easily adaptable for various more complicated multi-objective problems, such as multi-objective optimization of the final state of a dynamic model, when the requirements of repeating p times rather complex optimization calculations and of using attainable $\bar{y}$ are quite restrictive. Therefore, it is better to remove such requirements by using penalty methods when solving (8a,b) - see [54]. An application of an <u>external quadratic penalty function</u> leads to the following modification of (8a,b):

$$\tilde{h}_j^{\varrho}(\bar{y}) = \max_{y \in Y} \; s_j(y,\bar{y},\varrho) \tag{9a}$$

where:

$$s_j(y,\bar{y},\varrho) = y_j + 0.5\varrho \sum_{i \neq j} (\max(0, \bar{y}_i - y_i))^2 \tag{9b}$$

with some penalty coefficient $\varrho > 0$. We replace thus constraints with penalty terms, interpreted as soft constraints that might be violated; therefore, $\bar{y}$ need not be attainable. Moreover, if these constraints are all violated at a maximal point $\hat{y}$ of $s_j(.,\bar{y},\varrho)$, $\hat{y}_i < \bar{y}_i$ for all $i \neq j$, then $\hat{y} \in \hat{Y}^p$, it is properly Pareto-optimal - as a result of a single, not repeated optimization - because the function $s_j(.,\bar{y},\varsigma)$ is then strongly monotone and its maximal points correspond to stable problems (8a) with bounded Lagrange multipliers - see e.g. [55]. The penalty function would just stubbornly refuse to find a maximum at an improperly Pareto-optimal outcome, would find instead a properly Pareto-optimal outcome that is close to the improperly Pareto-optimal one.

2.4. Aspirations and achievement scalarizing functions.

The above property suggests another way of checking the stability of the problems (8a,b) and proper Pareto-optimality of related outcomes - even without convexity assumptions. We would use, however, exact nondifferentiable penalty functions, which is connected to another issue - how to construct, with help of a norm, a scalarizing function that would characterize parametrically properly efficient outcomes (with a prior bound) while using an aspiration or reference point $\bar{y}$ as a parameter not restricted to $\bar{y} \in \bar{Y}$ nor to $\bar{y} \in Y$ - in other words, how to remove the disadvantages of both displaced ideal and goal programming methods.

The construction of such a function is related to the concepts of the cone $C(\varepsilon)$ and of separating the interior of this cone from Y by a nonlinear function. If $C = \mathbb{R}^p_+$ and we use the l_1 norm on the right-hand side of the inequality in (6a) as well as the l_∞ norm augmented with l_1 to define $\text{dist}(y,C)$, then we obtain the cone $C(\varepsilon) = C(\varepsilon,l_1,l_\infty)$:

$$C(\varepsilon,l_1,l_\infty) = \{y \in \mathbb{R}^p : ||y^{(-)}||_{l_\infty} + 2\varepsilon||y^{(-)}||_{l_1} \leq \varepsilon||y^{(-)}||_{l_1}\} \quad (10a)$$

where:

$$y^{(-)} = (\min(0,y_1),\ldots\min(0,y_i),\ldots\min(0,y_p)) \quad (10b)$$

This cone can be equivalently written in several forms, such as:

$$C(\varepsilon,l_1,l_\infty) = \{y \in \mathbb{R}^p : -y_j \leq \varepsilon \sum_{i=1}^{p} y_i, \ j = 1,\ldots p\} =$$

$$= \{y \in \mathbb{R}^p : \min_{1 \leq i \leq p} y_i + \varepsilon \sum_{i=1}^{p} y_i \geq 0\} \quad (10c)$$

The first of these forms implies that $C(\varepsilon,l_1,l_\infty)$ is a convex polyhedral cone. The second indicates that the interior of this cone can be strictly separated from the rest of the space by a nonlinear and nondifferentiable, but continuous and monotone function; by appropriately shifting this function, we can also separate a shifted cone $\bar{y}$ + int C( ).

The concept of separation is very strong: the statement that a function strictly separates, at some point $\bar{y} \in Y$, the set Y and the shifted cone $\bar{y}$ + int C(ε), is equivalent to three statements: (i) that $\bar{y}$ maximizes this function over $y \in Y$; (ii) that $\bar{y} \in \tilde{Y}^{pe}$, it is properly efficient with prior bound; and (iii) that this function is C(ε)-strictly monotone (at least at the point $\bar{y}$). With help of this equivalence, it can be shown - see [45] - that the following scalarizing function:

$$s(y,w) = \min_{1<i<p} (y_i - \bar{y}_i) + \varepsilon \sum_{i=1}^{p} (y_i - \bar{y}_i); \quad w = \bar{y} \in W^s = \mathbb{R}^p \qquad (11a)$$

completely characterizes, without any convexity assumptions, properly Pareto-optimal outcomes with a priori bound ε - and that trade-off coefficients in this case are indeed bounded by $1 + 1/\varepsilon$.

Moreover, a single maximization of this function is equivalent to solving p times the following modified perturbation problems (8a,b):

$$\tilde{h}_j^\varepsilon(\bar{y}) = \max_{y \in \tilde{Y}^{(j)\varepsilon}(\bar{y})} (y_j + \varepsilon \sum_{i=1}^{p} y_i) \qquad (11b)$$

where:

$$\tilde{Y}^{(j)\varepsilon}(\bar{y}) = \{y \in Y: y_i \geq \bar{y}_i + \varepsilon \sum_{k=1}^{p} (\bar{y}_k - y_k) \text{ for all } i \neq j\} \qquad (11c)$$

Therefore, by maximizing the function (11a), we check also the stability of perturbation problems (8a,b) - which are limits of (11b,c) as $\varepsilon \to 0$. Naturally, the function (11a) and the problems (11b,c) are formulated assuming that all objective outcomes are earlier normalized - we do not use here the symbols y_n, $\underset{n}{y}$ just for notational simplicity.

The scalarizing function (11a) is an example of a much broader class of order-consistent achievement scalarizing functions, see [44], with level sets that tightly approximate the cones $\bar{y}$ + C or $\bar{y}$ + C(ε). Such functions are basic for aspiration-led decision support, since they are parameterized by aspiration, reservation or reference levels without requiring that these parameters must be attainable or not attainable.

The main advantage of achievement scalarizing functions is that they typically result in locally parametric controllable characterizations and that they completely characterize properly efficient solutions with prior bound also for nonconvex problems; therefore, they are broadly applicable - to mixed integer optimization problems, to time discrete dynamic models, etc. There exist many variants of such functions - see [28] - that are devised for joint use of reservation and aspiration levels or are adapted for dynamic models. Usually, they are nondifferentiable functions, which is a direct consequence of the property of approximating the pointed cones $\tilde{y}$ + C and a price paid for a characterization that is independent of convexity assumptions. Therefore, the general problem of maximizing such an achievement function with a given substantive model of a decision situation must be solved by nondifferentiable optimization techniques. However, if the model is of linear programming type, the maximization problem can be rewritten as a linear programming problem; and for nonlinear differentiable models, differentiable approximations of achievement functions can be also constructed, see [44], [28].

2.5. Extensions: dynamics, uncertainty, game-like decision situations.

There might be diverse understanding of dynamic aspects of multi-objective optimization and decision support - see [56]. An optimization model of dynamic nature can lead either to multi-objective optimization of the final state, or sequential multi-objective optimization of dynamic programming type, or multi-objective optimization of trajectories. If we consider various uncertainty issues, then their probabilistic, or fuzzy set, or set-valued models are deeply related to multi-objective optimization and usually have also their dynamic aspects. On the other hand, a decision process is also multi-objective and dynamic in its essence, in all its main phases of intelligence (data acquisition), design (problem formulation) and choice (of actual decision). Especially important are dynamic aspects of learning and changing preferences during a decision process.

In order to give an example of some of these issues, we shall briefly consider a dynamic model described by a difference equation:

$$x(t+1) = \tilde{f}(x(t),u(t),t), \quad t = t_0,\ldots t_f; \quad x(t_0) = x_0 \in \mathbb{R}^n \qquad (12a)$$

where $x(t) \in \mathbb{R}^n$ is the state variable, $u(t) \in U \subset \mathbb{R}^m$ - the control variable, U being an admissible control set. THe continuity of the function $\tilde{f}: \mathbb{R}^n \times \mathbb{R}^m \times \mathbb{R}^1 \to \mathbb{R}^n$ is sufficient for the existence of solutions of (12a) on any finite discrete time interval $[t_0,\ldots t_f]$, provided U is bounded; but we usually assume differentiability in order to apply gradient-like optimization techniques. The set of admissible controls might be defined by additional inequalities of the type $g(u(t)) \leq 0$; admissible control sequences $u = (u(t_0),...u(t_f-1))$ - which can be identified with admissible decisions - belong to the space of bounded functions from the discrete interval $[t_0,...t_f-1]$ into $\mathbb{R}^m$.

The decision outcomes related to the model (12a) can be defined in various ways; in a simplest case, we can consider the optimization of the final state, where the outcomes y are defined by the equation:

$$y = h(x(t_f)) \tag{12b}$$

where $h: \mathbb{R}^n \rightarrow \mathbb{R}^p$ is a continuous (or differentiable) function; the sense of optimality in the outcome space can be again defined by a positive cone, say $C = \mathbb{R}^p_+$. We could also assume that y depends on all previous $x(t)$, $u(t)$ for $t = t_0,...t_f-1$, but such a dependence can be rewritten to the form (12b) through a suitably extended definition of state variable. Given a control sequence $u = (u(t_0),...u(t_f-1))$, any solution $x(\tau)$ of (12a) can be expressed as a transition function:

$$x(\tau) = X(\tau,t,x(t),u[t,...\tau-1]) \tag{12c}$$

where $\tau \geq t + 1$ and $u[t,...\tau-1]$ denotes the restriction of the control sequence to the discrete subinterval $[t,...\tau-1]$. With $t = t_0$, $\tau = t_f$ we obtain:

$$y = h(X(t_f,t_0,x_0,u)) = f(u) \tag{12d}$$

which can be interpreted that the dynamic model defines, after all, a static dependence of y on every component of u. However, dynamic models do have their specific features that must be taken into account when devising standards of formulation of such models, can simplify computing gradients of outcome functions through symbolic structural gradient manipulation, or do influence the convexity of outcome functions. For example, for continuous-time dynamic models one can show the convexity of the set Y of attainable outcomes if h is a linear function, no matter what the form of $\tilde{f}$ and U, see e.g. [56] - while for discrete-time dynamic models of the type (12a,b) we can be sure of convexity of this set only if the functions $\tilde{f}$, g, h are linear and the set U is convex.

On the other hand, we should not forget about the equivalence to a static problem and use it whenever it simplifies reasoning or computations. This is pertinent, for example, in relation to multi-objective sequential optimization and dynamic programming research, a subject studied recently rather intensively - see [57], [58]. The transition property (12c) is known to be essential for the optimality principle and the resulting dynamic programming technique in single-objective dynamic optimization. The question is whether this principle and technique can be extended to the multi-objective case.

The answer is positive under additional assumptions that can be stated in various ways, but generally relate to separability and monotonicity. The question of a most natural form of such assumptions is still open, see [56], [57]. This is an important theoretical issue, but its practical importance is rather limited, as indicated by the equivalence to a static problem and the following comments.

Historically, dynamic optimization is a much older and broader subject than the particular technique called dynamic programming as proposed by S.E. Dreyfus and further substantially extended and popularized by R.E. Bellman. This particular technique is very powerful theoretically but not necessarily effective as a computational tool. Specifically, it is a good computational tool for special problems with a small, discrete number of admissible states of a dynamic process, but a very poor computational tool whenever the number of admissible states is large - in particular, if admissible states belong to a set of continuum power, as in the model (12a,b). When speaking about "the curse of dimensionality" - that is, the exponential dependence of computational effort on the dimensionality of state space - Bellman did not mention that this is a feature of only rather simplistic approaches to dynamic optimization that he used as a comparison basis for the dynamic programming technique (which also displays an exponential, though slower than the simplistic techniques, growth of computational effort with the dimensionality of state space).

But other techniques of dynamic optimization, though they are often less general than dynamic programming, are in fact much more effective computationally. For example, if all functions in the model (12a,b) are linear and a piece-wise linear scalarizing function such as (11a) is used, then its maximization is equivalent to a dynamic linear programming problem. Computational effort when solving such a problem can be shown today to have only polynomial dependence on the dimensionality of state space (due to the results of N. Shor, L. Khachian, L. Karmarkar and others, made known through the celebrated though generally rather misinformed dispute on the computational complexity).

Dynamic linear programming techniques can be also effectively applied to the problems of multi-objective trajectory optimization, obtained e.g. if we consider the model (12a,b) but assume that decision outcomes are defined for each time instant $t = t_0, \ldots t_f$ - see [56], [28]. The essential advantage of dealing with entire trajectories of outcomes is that human mind can interpret and evaluate them holistically, by "Gestalt" - which makes it much easier to specify, for example, reference or aspiration trajectories for selected objective components. While we have difficulties when dealing in our minds with more than seven or nine objects at once, this does not mean that these objects must be necessarily single numbers; they might as well correspond to any familiar concept, shape or profile - or trajectory - which, when treated analytically, might be even described as an element of an infinite-dimensional space and in a computer would be represented by a large number of similar data, but in our minds becomes aggregated. Thus, it is advisable to group into outcome trajectories (over time) or profiles (over space or other variables) any larger number of similar outcomes, not necessarily in dynamic models.

The questions of modeling uncertainty are very broad and cannot be covered in a short outline. We shall stress here only their relation to multi-objective and dynamic optimization. The traditional, probabilistic form of uncertainty models has led to problems of stochastic optimization - but the concepts of stochastic optimization with recourse or sequential stochastic optimization are in their essence both multi-objective and dynamic, see e.g. [59], [60]. The same can be said about the related computational techniques - see e.g. [61].

Fuzzy set models of uncertainty have been developed in close relation to multi-objective decision making and optimization, see e.g. [4]. Set-valued models of uncertainty have been also essentially connected with dynamic and multi-objective optimization - see [62]. A general challenge relates to computational techniques for models with uncertainty: such techniques must be sufficiently powerful to be applicable to repetitive, parameterized optimization in decision support systems.

The issue of learning in a decision process has been long ago - see [63] - perceived as essentially related to that of uncertainty - it is actually impossible to learn without mistakes or at least a noise. As already mentioned, more recent studies of the mechanism of expert decision making underline the qualitative change of decisions as the basic result of learning, see [10]. In order to include the dynamics of learning into analytical approaches to decision support, further mathematical formalization of the problem of learning and expert decision making is necessary. With a few exceptions - see e.g. [64] - there is little attention paid to this important problem.

A different and rich set of issues related to group decision making and game-like decision situations can be only outlined here. Many decision makers in a group, even if they do not have the same objectives, might be prevented from making independent decisions by institutional constraints - or even by physical impossibility of implementing more than one decision (such as a family travelling in a car and discussing which route to follow). On the other hand, even if they would have the same objectives and preferences, they might have different information - which makes necessary to come to a joint decision through a discussion. The situation in which a group discusses a joint decision is called group decision making; the subject of decision support for group decision making - see e.g. [33] - is studied today rather intensively.

A game-like decision situation differs from group decision making by the possibility of actual implementation of independent decisions by each decision maker (actor, player). For example, if members in a group voluntarily agreed to accept joint decisions, they can threaten to defect and change the group decision making into a game-like situation. Many results from game and bargaining theory have been applied to studying game-like decision situations, see e.g. [65]. However, most of these results assume either that players have single objectives or that their multiple objectives are aggregated by known and unchanging utility functions. Since such assumptions are impractical, other researchers conclude that game theory is not applicable for analyzing game-like decision situations and prefer gaming simulation; but even for simulated gaming it is necessary to develop decision support, see [66]. Therefore, a much needed but scarcely represented - see [67], [45] - direction of research is related to multi-objective bargaining and game theory.

3. OTHER METHODOLOGICAL ISSUES AND CONCLUSIONS.

There are many methodological issues that could not be fully discussed in the above outline of the theory and methodology of aspiration-led decision support. We can approach them, for example, by postulating that consecutive phases of decision support should correspond to all phases of a decision process. H. Simon [12] divides a decision process into the phases of intelligence, design and choice. S. Cooke and N. Slack [68] distinguish much more phases of a problem solving and decision making process, including observation, a formal recognition of a problem, interpretation and diagnosis, the definition of a decision problem, the determination of options or alternatives, the evaluation of options, selection and choice, finally decision implementation and monitoring. The experience in designing ALDSS indicates that this list is not complete yet. After interpretation and diagnosis, several steps of modeling the problem can accompany attempts to define the problem more precisely; additional data might be gathered, various types of analytical or logical models might be perused, diverse definitions of the problem - including various objectives and various ways of aggregating them - might be tested on the models. The decision process might have many recourse loops; even if a preliminary decision has been made, its post-optimal sensitivity analysis might lead to a new problem formulation. All these phases can have specific type of computerized support; we shall comment her shortly upon some related issues.

An important issue is the relation between decision analysis and support on one side and data gathering and model building as an initial step in decision analysis on the other side. There exist software packages or even programming languages (such as RATS or GAUSS) that support diverse techniques of statistical parameter estimation for model building. However, their final aim is to produce a model - and not to use this model in decision support. Therefore, they do not - for example - address the question what is the meaning for a better model for a particular decision situation - a meaning that usually does not correspond - see [55] - to the general but not necessarily constructive concept of a better statistical fit.

Another important issue is that of standards of model programming, editing and simulation. There exist logical programming languages as well as simulation languages for models of analytical type; however, these languages were not developed for the needs of decision support systems. After accumulating more experience in DSS design, a new development of such tools will be necessary. Meanwhile, it is important to choose standards for model programming and editing that would be useful not only for model simulation but also for their multi-objective analysis and optimization. Such standards exist (e.g. MPSX) only for linear programming models and even then are not best suited for multi-objective analysis. For nonlinear models, an essential element of such standards should be an automatic, symbolic differentiation of all formulae, taking into account their structural relations in a model - see [69].

A related issue is that of optimization tools. An impressive development of optimization algorithms during last forty years has its side-effects: algorithm development often becomes a goal in itself - and such algorithms require considerable tuning of their parameters, they can work effectively only when tuned for a given class of problems by their author. Such algorithms are of little use in DSS development where we need robust optimization algorithms that would work reliably for a broad class of problems without parameter adjustment. Robust optimization algorithms exist for linear programming problems (including algorithms of non-simplex type that were known long before Khachian and Karmarkar, see [70]) and for differentiable nonlinear optimization, see [69]; but it is important to develop them for nondifferentiable and mixed continuous-discrete (combinatorial) optimization.

Other sets of issues relate to interactive graphics for DSS, to post-optimal sensitivity analysis, to combining analytical optimization with logical inference modeling in DSS, etc. Most of these issues require further studies and are promising subjects for future research. But the experience already gathered in the development of ALDSS is considerable, proving that it is practical to construct decision support systems based on multi-objective analysis and optimization.

REFERENCES.

1. Sawaragi, Y., H. Nakayama and T. Tanino: Theory of Multiobjective Optimization. Academic Press, Orlando Fl., 1985.
2. Yu, P.L.: Multiple-Criteria Decision Making - Concepts, Techniques and Extensions. Plenum Press, New York and London 1985.
3. Steuer, R.E.: Multiple Criteria Optimization: Theory, Computation and Application. J. Wiley, New York 1986.
4. Seo, F. and M. Sakawa: Multiple Criteria Decision Analysis in Regional Planning: Concepts, Methods and Applications. D. Reidel Publishing Company, Dordrecht 1988.
5. Keen, P.G.W. and M.S. Scott Morton: Decision Support Systems - an Organizational Perspective. Addison - Wesley, 1978.
6. Bonczek, R.H., C.W. Holsapple and A.B. Whinston: Foundations of Decision Support Systems. Academic Press, New York 1981.
7. Wierzbicki, A.P.: The use of reference objectives in multiobjective optimization. In G. Fandel and T. Gal, eds.: Multiple Criteria Decision Making, Theory and Applications. Springer Verlag, Heidelberg, 1980.
8. Wierzbicki, A.P.: A mathematical basis for satisficing decision making. Mathematical Modeling, Vol. 3 (1982) pp. 391-405.
9. Grauer, M., M. Thompson and A.P. Wierzbicki, eds.: Plural Rationality and Interactive Decision Processes. Proceedings, Sopron 1983, Hungary. Springer Verlag, Berlin - Heidelberg, 1984.
10. Dreyfus, S.E.: Beyond rationality. In [9], 1984.

11. Simon, H.: Models of Man. MacMillan, New York 1957.
12. Simon, H.: Administrative Behavior. MacMillan, New York 1958.
13. Tietz, R., W. Albers and R. Selten, eds.: Bounded Rational Behavior in Experimental Games and Markets. Proceedings, Bielefeld 1986. Springer Verlag, Berlin - Heidelberg, 1988.
14. Galbraith, J.K.: The New Industrial State. Houghton - Mifflin, Boston 1967.
15. Rappoport, A.: Uses of experimental games. In [9], 1984.
16. Axelrod, R.: The Evolution of Cooperation. Basic Books New York 1985.
17. Charnes, A. and W.W. Cooper: Goal programming and multiple objective optimization. J. Oper. Res. Soc., Vol. 1 (1977) pp. 39-54.
18. Ignizio, J.P.: Goal programming - a tool for multiobjective analysis. Journal for Operational Research, Vol. 29 (1978) pp. 1109-1119.
19. Glushkov, V.M.: Basic principles of automation in organizational management systems (in Russian). Upravlayushcheye Sistemy i Mashiny, Vol. 1 (1972).
20. Pospelov, G.S. and V.A. Irikov: Program- and Goal-Oriented Planning and Management (in Russian). Sovietskoye Radio, Moscow 1976.
21. Umpleby, S.A.: A group process approach to organizational change. In H. Wedde, ed.: Adequate Modeling of Systems. Springer Verlag, Berlin - Heidelberg 1983.
22. Wierzbicki, A.P.: Negotiation and mediation in conflicts, I: The role mathematical approaches and methods (in H. Chestnut et al., eds.: Supplemental Ways to Increase International Stability, Pergamon Press, Oxford 1983), II: Plural rationality and interactive decision processes (in [9] 1984).
23. Haimes, Y.Y. and W.A. Hall: Multiobjectives in water resource systems analysis: the surrogate trade-off method. Water Resource Research, Vol. 10 (1974) pp. 615-624.
24. Larichev, O.I.: Man-machine procedures for decision making. Automation and Remote Control, Vol. 32 (1972) pp. 1973-1983.
25. Fandel, G.: Optimale Entscheidungen bei mehrfacher Zielsetzung. Springer Verlag, Heidelberg 1972.
26. Keeney, R. and H. Raiffa: Decisions with Multiple Objectives: Preferences and Value Trade-Offs. J. Wiley, New York 1976.
27. Kallio, M., A. Lewandowski and W. Orchard-Hays: An implementation of the reference point approach for multi-objective optimization. WP-80-35, IIASA, Laxenburg 1980.
28. Lewandowski, A. and A.P. Wierzbicki, eds.: Aspiration Based Decision Support Systems. Springer Verlag, Berlin - Heidelberg 1989.
29. Grauer, M., A. Lewandowski and L. Schrattenholzer: Use of the

reference level approach for the generation of efficient energy supply strategies. WP-82-19, IIASA, Laxenburg 1982.

30. Grauer, M. and E. Zalai: A reference point approach to nonlinear macroeconomic planning. WP-82-134, IIASA, Laxenburg 1982.

31. Messner, S.: Natural gas trade in Europe and interactive decision analysis. In G. Fandel et al., eds.: Large Scale Modeling and Interactive Decision Analysis. Springer Verlag, Berlin - Heidelberg 1985.

32. Strubegger, M.: An approach for integrated energy - economy decision analysis: the case of Austria.. In G. Fandel et al., eds. (see [31]).

33. Lewandowski, A., S. Johnson and A.P. Wierzbicki: A selection committee decision support system. In Y. Sawaragi, K. Inue and H. Nakayama, eds.: Towards Interactive and Intelligent Decision Support Systems, Springer Verlag, Berlin - Heidelberg 1986.

34. Nakayama, H. and Y. Sawaragi: Satisficing trade-off method for multi-objective programming. In M. Grauer and A.P. Wierzbicki, eds.: Interactive Decision Analysis, Springer Verlag, Berlin - Heidelberg 1983.

35. Nakayama, H.: Sensitivity and trade-off analysis in multiobjective programming. In [39].

36. Steuer, R.E. and E.V. Choo: an interactive weighted Chebyshev procedure for multiple objective programming. Mathematical Programming Vol. 26 (1983) pp.326-344.

37. Korhonen, P. and J. Laakso: Solving a generalized goal programming problem using a visual interactive approach. European Journal of Operational Research, Vol. 26 (1986), pp. 355-363.

38. Korhonen, P. and J. Wallenius: A careful look at efficiency in multiple objective linear programming. In A.G. Lockett and G. Islei, eds.: Improving Decision Making in Organisations, Proceedings, Manchester 1988. Springer Verlag, Berlin - Heidelberg 1989.

39. Lewandowski, A. and I. Stanchev, eds.: Methodology and Software for Interactive Decision Support. Springer Verlag, Berlin - Heidelberg, 1989.

40. Kuhn, H.W. and A.W. Tucker: Nonlinear Programming. In Proceedings of Second Berkeley Symposium on Mathematical Statistics and Probability, University of California Press, Berkeley, Cal. (1950), pp. 481-492.

41. Geoffrion, A.M.: Proper efficiency and the theory of vector optimization. Journal of Mathematical Analysis and Applications, Vol 22 (1968), pp. 618-630.

42. Henig, M.I.: Proper efficiency with respect to cones. Journal of Optimization Theory and Applications, Vol. 36 (1982) pp. 387-407.

43. Wierzbicki, A.P.: Basic properties of scalarizing functionals for multiobjective optimization. Mathematische Operationsforschung und Statistik, s. Optimization, Vol. 8 (1977) pp. 55-60.

44. Wierzbicki, A.P.: On the completeness and constructiveness of parametric characterizations to vector optimization problems. OR-Spektrum, Vol. 8 (1986) pp. 73-87.

45. Wierzbicki, A.P.: Multiple criteria solutions in noncooperative game theory. Part III: Theoretical foundations. Kyoto Institute of Economic Research, Discussion Paper No. 288, 1990.

46. Salukvadze, M.L.: Vector Valued Optimization Problems in Control Theory. Academic Press, New York 1979.

47. Volkovich, V.L.: Multicriteria problems and methods of their solutions (in Ukrainian). In Complex Control Problems, Izdatelstvo Naukova Dumka. Kiev 1969.

48. Salukvadze, M.L.: On the optimization of vector functionals, 1. Programming optimal trajectories, 2. Analytic construction of optimal controllers. Automation and Remote Control, Vol. 31 (1971) No. 7, 8.

49. Yu, P.L. and G. Leitmann: Compromise solutions, domination structures and Salukvadze's solution. JOTA Vol. 13 (1974), pp. 362-378.

50. Zeleny, M.: Compromise programming. In J.L. Cochrane and M. Zeleny, eds.: Multiple Criteria Decision Making. Univ. of South Carolina Press, Columbia S. Carolina 1973.

51. Zeleny, M.: Multiple Criteria Decision Making. McGraw Hill, New York 1982.

52. Geoffrion, A.M.: Duality in nonlinear programming: a simplified application-oriented development. SIAM Review Vol. 13 (1971) pp. 1-37.

53. Benson, H.P. and T.L. Morin: THe vector maximization problem: proper efficiency and stability.SIAM Journal on Applied Mathematics, Vol. 32 (1977) pp. 64-72.

54. Wierzbicki, A.P.: Penalty methods in solving optimization problems with vector performance criteria. Proceedings of the Vi-th IFAC World Congress, Cambridge Mass. 1975.

55. Wierzbicki, A.P.: Models and Sensitivity of Control Systems. Elsevier, Amsterdam 1984.

56. Wierzbicki, A.P. Dynamics aspects of multi-objective optimization. Proceedings of Yalta Conference on Vector Optimization, Springer Verlag, Heidelberg - Berlin 1990.

57. Li, D. and Y.Y. Haimes: The envelope approach for multiobjective optimization problems. IEEE-SMC, Vol.17 (1987) pp. 1026-1038.

58. Li, D. and Y.Y. Haimes: Multiobjective dynamic programming - the state of the art. In A.G. Lockett and G. Islei, eds. (see [38]).

59. Wets, R.J.B.: Stochastic programming: solution schemes and approximation techniques. In A. Bachem et al., eds.: Mathematical Programming: The State of the Art. Springer Verlag, Berlin - Heidelberg 1983.

60. Ermolev, Yu.M. and R.J.B. Wets: Numerical Methods in Stochastic Programming. Springer Verlag, Berlin - Heidelberg 1987.

61. Ruszczynski, A.: Modern techniques for linear dynamic and stochastic programs. In [28].

62. Kurzhanski, A.B.: Inverse problems in multiobjective stochastic optimization. In Y. Sawaragi et al., eds. (see [33]).

63. Feldbaum, A.A.: Foundations of the Theory of Optimal Control Systems (in Russian). Nauka, Moscow 1962.

64. Michalevich, M.V.: Stochastic approaches to interactive multicriteria optimization problems. WP-86-10, IIASA, Laxenburg 1986.

65. Roth, A.E.: Axiomatic Models of Bargaining. Springer Verlag, Berlin - Heidelberg 1979.

66. Wierzbicki, A.P.: Multiobjective decision support for simulated gaming. In A.G. Lockett and G. Islei, eds. (see [38]).

67. Bronisz, P., L. Krus and A.P. Wierzbicki: Towards interactive solutions in a bargaining problem. In [28].

68. Cooke, S. and N. Slack: Making Management Decisions. Prentice - Hall, Englewood Cliffs 1984.

69. Kreglewski, T., J. Paczynski, J. Granat and A.P. Wierzbicki: IAC-DIDAS-N - a dynamic interactive decision analysis and support system for multi-criteria analysis of nonlinear models. In [28].

70. Makowski, M. and J.P. Sosnowski: Mathematical programming package HYBRID. In [28].

DSS MIDA: LESSONS FROM EXPERIENCE IN DEVELOPMENT AND APPLICATION IN THE CHEMICAL INDUSTRY

M. Zebrowski

Academy of Mining and Metallurgy, Cracow, Poland

ABSTRACT

Paper aimed at documenting a personal view of the subject as given by author in Udine Summer School, presents his experience gained from development and extensive application of a DSS MIDA — Multiobjective Interactive Decision Aid. Using a particular DSS presented in its broader environment of applications namely in the Integrated Development Programming (IDP), an attempt is made to draw observations from identification, modeling and structuring of the particular DSS in order to share more generally applicable experience. This concerns counter-intuitive theoretical solution applied in decomposition of the multiobjective dynamic decision problem of industrial development. Lessons from variety of applications cathegorized depending on the environment, scope and relation to the user are summarized in order to show some more general experience. Hopefully, a kind of practically useful philosophy of DSS development and application is presented through the assumed scope.

1 Some personal remarks

Summer Schools especially such as the one in Udine, offer a unique opportunity. Speaking in my own name I dare to consider it an opportunity to all participants i.e. those who come to receive some knowledge and those who are supposed to deliver it. In order to assure and establish proper communications and exchange of views, attitudes and experience, each discussed topic or subject in question should be conveyed in a way both sufficiently general and appealing to the people who are interested in, say, decision support but do not necessarily have experience or interest in the particular DSS or particular field of application. At the same time, the essential technicalities must be also presented to provide formal and factual base or at least a clue.

As opposed to the above it is definitely not the case of typical scientific meetings where generalization, wholistic interpretation, philosophy and experience are almost as a rule sacrificed for the benefit of purely scientific (whatever that means), technical, narrowed to the particular field and even case considered presentation. It usually makes understanding of such a presentation limited to a very narrow group of specialists. It proves to be very difficult however to really take advantage of the opportunity offered by the school since in fact it only lasts during the event through direct contacts between people in a particular place. Therefore delivering a paper which would convey the experience in a way appropriate to such a unique occasion proves to be much more difficult. Sharing ideas and experience gained from complete development cycle of a DSS and a variety of its applications covering period of almost 15 years, I am fully aware how difficult if not impossible is such an attempt. I would like to conclude this personal kind of remarks with the message which I find perhaps the most important experience coming from the relatively long practising in DSS development and application: how important it is to sustain modesty in assessing results as well as in sharing advice when dealing with such complex phenomena as those which occur in decision environment especially in industrial development programming. I am sure that this message as well as a kind of carefully balanced mixture of courage and caution which have to be sustained in decision support will be shared by all experienced practicioners of the field.

2 What is it about

I would like to present here some experience gained from efforts invested in theory, software and applications in the development of a DSS such as MIDA. MIDA or — Multiobjective Interactive Decision Aid is a decision support system developed and applied for so called Integrated Development Programming (IDP) basically in refining, petrochemical and chemical industries.

To cover the subject following scope is assumed. Prior to characteristics of the field of application is given showing how the MIDA system and methodology emerged from the identification process, some remarks on philosophy of modeling are given. Then the essence of Integrated Development Programming or IDP is explained. Some basic categorization of MIDA applications shows flexibility and generality of the DSS.

This yields also a proper base for some concise considerations on the architecture of the system. The basic model which provides a theoretical framework and consequently a core of the software developed is given in the Appendix. Some final remarks close the assumed scope.

3 On philosophy of modeling

Science, management and art

One is to be aware that the type of modeling which is a core of programming industrial development as understood here should be regarded as combined activities of science, management and art. Therefore no strict rules can be devised and observed in order to obtain "the best solution". That is the reason why we propose here a term philosophy of modeling as opposed to methodology.

We are dealing with the class of systems called socio-economic systems. The reader is encouraged to refer as strongly as possible to his own not only professional but also life experience. The goal would be to merge the knowledge from this paper with own experience which should enable to work out an individual creative approach to problems not necessarily limited to development programming.

In the subtitle above we have explicitly used the term modeling which means that a model must become the center of interest in the process. Putting it briefly, modeling is problem solving based on the simplified description (i.e. a model) of the nature of system under consideration. The term simplification is very important perhaps even the most important, since the simplification has to be done from the point of view of the problem to be solved.

Two elements must be present in order to talk about a model and modeling, namely the system object or phenomenon, and a problem which sets up the point of view and the goal for modeling.

The models built for forecasting purposes make no exception to this rule: the difference is not real but only apparent. There must be some existing structure and its behavior must be clear in order to find out about its possible future (i.e. forecasted) state. Continuity must be preserved between the present and the future state. Having a problem and consequently a model based on such initial conditions we have to simulate along various scenarios in order to find out the changes acceptable, first of all, by the existing structure and only then we can judge the forecast's acceptability. It is also not uncommon case in socio-economic systems that we are very often working out a forecast in order to find how to avoid its fulfillment (it is even called a warning forecast).

Without going into the general discussion on problem solving sequence we should comment on the meaning of an expression used here: "to solve a problem". By saying "to solve a problem" we generally mean that, first: we achieve the understanding of the phenomenon and second: we know what to propose to be undertaken as an action. The lack of proposed action may be only apparent or it can be suspended temporarily for the next phase.

The very important remark should be formulated here, which is a part of the philosophy of modeling, is that **the choice of a particular model and a particular type of an analysis and simulation technique means, in fact, a decision of excluding all other possibilities.** The art of the problem solving is strongly dependent on the consciousness of this fact.

When we are talking about a solution in terms of techniques we deal again with rather apparent, not real, problem of relations between the simulation and the optimal solution.

It should be pointed out, however, that at least **in the socio-economic systems there** is no optimal solution to a problem, since the "optimal solution" is limited only to the assumed conditions in the environment and the identified (not the same as real) internal structure of the system. **Therefore, we can only simulate the solution which is optimal under a given set of conditions.** This is not merely an academic problem since some modelers tend to lobby for the optimal solution. The optimal solution should be considered and used only as a technical term for a specific, formalized problem.

Model: problem solving and solving system

If we recall now a basic principle of modeling, i.e. the ability to build an image being a purposefully simplified picture of reality, then we have to face a very important consequence of this principle which may be called a rule. The rule **stems from the fact that the model has to contain the minimum information necessary for solving the problem and it has to produce the minimum information necessary for describing the solution.** Both **the incoming information** and, even to a larger degree, **the outgoing information have to be digestible and manageable by a modeler.**

But how can this rule be fulfilled in the case of such complex systems as socio-economic ones? The models contain a great number of variables and parameters and are in fact, more and more complex. Even such models, however, can still remain simple and understandable provided that **one has tools enabling for an easy manipulation and handling all the data and information through all the stages of the problem solving.**

From this a concept of a **solving system** naturally evolves, **which means not only the pure computer technology but all the necessary organizational or technical tools which have to be accessed in the process of problem solving.** The information technology and the whole range of very diversive methods and tools are to be extensively used in order to fulfill the rule displayed above.

The importance of devising the solving system will be always differently judged depending on the type and scale of the problem that is to be solved. **Those who deal with the problems** such as in our case the **industrial strategy design**, will certainly agree on **the importance of designing the whole system of the data supply and storage, the data verification and processing as well as the manipulation and storing of the results.** It would be obvious for them that **the modeling (and models) is the part of the decision making system.** Some others dealing with the relatively isolated environmental problems will be certainly less concerned with the matter. But still they should not take a risk and neglect the problem since the difference is more apparent than real and all the above aspects must be also dealt with in a way specific for environmental problems.

Tower of Babel and tooler's effects

The ever growing diversity of methods and techniques available leads to another trap. In fact all the phenomena (problems) which one can find in the socio-economic systems (food, agriculture, industry, energy, population problems) are very complex in themselves. Besides they interact strongly which is one of the main difficulties in solving them. In real life all the phenomena of concern are present and interlinked in the socio-economic systems. That means that "they use some common language". But when we come to the modeling stage, to the methods and the corresponding solving systems, this communication is lost. First it happens at the stage of setting up the systems boundaries and then the gap grows through the stages of setting corresponding solving systems and application of the particular modeling methods and simulation techniques. What is worse, it affects also modelers – those who specialize in solving different problems (dealing with specific phenomena) cannot communicate among themselves. This syndrome can be called a **"Tower of Babel effect"**. In all we may summarize: the consciousness of this danger and ability to overcome it is a part of the philosophy. In fact the secret of non-existence of the true interdisciplinary approach lies in modeling of socio-economic systems.

To complete this set of problems let us regard one more, closely related to the above phenomena. It may be called a **"tooler's effect"** which describes another danger stemming from the great diversity of modeling techniques. Modelers become tool-oriented rather than problem-oriented. Such an attitude turns them against the fundamental principle of the modeling, which should be regarded as a synonym of the problem solving. On the contrary a tooler tends to apply a specific tool rather than to solve a real problem.

Perhaps we should end here the general considerations by the following conclusion. The search for philosophy of modeling should be a permanent task of anyone who deals with the modeling of socio-economic and complex technological systems and this can be achieved only by experience. **It means that the best philosophy is the maximum experience.**

4   Scope and perspective of development programming

Chemical industry and management of change

The world is permanently passing through a chain of great economic, social and technological changes. Recognition of this fact and of the need to control the forces of change has stimulated world-wide interest in the problems of change and methods for coping with them.

Nowhere is the need for management of change more crucial than in the industrial sector, where many factors can affect the growth or decline of individual industries and the resulting industrial structure. The process of change with perhaps the highest impact affects the chemical industry.

Here we concentrate on management of the chemical industry due to the problems it faces as a result of global change, particularly as a result of changing patterns of raw materials and energy use.

The importance of the chemical industry is often greatly underestimated. Not only does it provide soaps, detergents, medicines, but also pesticides, fertilizers, synthetic rubbers, plastics, synthetic fibers... — in fact, our modern technological society could be said to be founded on the chemical industry.

One of the most surprising facts about this industry is that a large proportion of its many products are derived from only a very small number of starting materials, of which hydrocarbons are probably the most important.

As the processing of natural resources with mineral or agricultural origin proceeds, the chains are branching with each processing phase from one generation of intermediates to another. The developed chemical industry presents in fact ever growing network of interlinked technologies. Final or market goods originating from this network provide only a small share of the total chemical production which does not exceed 25% of the final turnover of the industry. The necessity to meet the challenge of the management of change led to a practical action it is to development of a a methodology capable of design of feasible restructuring and/or structuring alternatives in various sectors of the chemical industry (see e.g.[1], [5]).

The approach chosen takes into account a variety of interrelated and alternative production processes (either in use or under development), compares their efficiency, their consumption of different resources etc. and finds a combination of technologies that best meets assessed needs while staying within the limits imposed by the availability of resources and environmental constraints.

Programming development — MIDA approach

From the above essential overview two spheres to be identified emerge ([7]). First is the sphere of the present and forecast performance of the industry which is a result of the identification. It should be described formally in order to represent the changes that are to transform the industrial structure in time. Second is the sphere of management of the changes where decisions are to be worked out and a decision support is to be developed and naturally embedded into the decision process.

From the above follows that a basic model of industrial structure (IS) must be provided to map the first sphere. It must be based on important assumptions regarding the decision on the appropriate aggregation level as well as on the boundaries of the industrial structure considered. If the management of change is going to be executed through Integrated Development Programming, then such a sphere of activities can be considered as a process of design of an Industrial Development Strategy (IDS).

IDS design is considered as a decision process based on generation of efficient development alternatives expressed in terms of goals, critical or indispensable resources as well as selected array of technologies which are to be utilized. The alternatives are to be generated, selected and ranked along assumed efficiency measures.

The aim of the development programming or IDS design is a selection of the alternative which is to change the industrial structure by means of investment over the time. Due to the dynamic properties of the development process, and specifically the development cycle of technology ([3]), the time span under consideration is of the range of 10 – 15 years. The straightforward conclusion is that due to the dynamic nature of the development process, IDS design is to be treated and solved as a dynamic problem. It has to be strongly underlined, however, that any attempt to formulate a general multidimensional dynamic problem as a means for generating feasible development alternatives must lead to oversimplification and severe loss of important factors which should not be overlooked. At the same time, any decomposition must assure that through a coherent methodology all the subproblems can be solved as integral parts of the same system. A fundamental premise for the phenomena of development is the fact that as time perspectives become longer (5,10,15... years), the reliability and accuracy of data describing the future decreases.

To meet the challenge of a real application in a complex decision environment, a method better responding to a managerial practice was elaborated. It is based on a decomposition of this in fact dynamic problem along space and time. Before going into discussion of the kind of properties of a decision problem (or problems) that are to be formulated and solved, let us make a step further in the identification of IDS design.

To perform IDS design with focus on generation of development alternatives and their selection, **following elements are to be considered**:

- **Existing industrial structure** in terms of consumption coefficients capacities and relevant economic data,

- **Potentially available technologies** for construction of new plants,

The above two categories form a technological repertoire out of which a new industrial structure is to be devised providing that a harmony between existing elements and the new ones must be sustained. The next category of elements for analysis are:

- resources which are to be utilized in order to implement a new structure, such as investment, manpower, water, energy etc as well as resources which are to be supplied as feedstock to run the new structure.

- **some of the resources considered are** selected in a special way and **called critical resources due to the fact that their availability is a necessary condition** to make development alternative feasible.

Critical are those resources which are nominated as such by the decision maker for either being particularly scarce or difficult to obtain; examples may be crude oil, manpower, energy or capital. In practice **the set of critical resources is closely related to the set of criteria, since the aim is to find an optimal solution with respect to all critical resources.** Technological constraints are quite easily identified and are related to factors such as production capacities and operating conditions. All other elements in the analysis, such as demand for a particular product, the availability of (noncritical) raw materials, fall into the category of complementary or auxiliary information which describes environment to the industrial activities such as terms of trade — specifically prices. On the contrary, a demand for selected significant products as well as availability of selected significant feedstock falls into the category of critical resources.

It is clear that whether a resource is selected as critical or not, it depends on the formulation of the decision problem. In fact, a resource can be nominated to one category or to the other by the decision maker and that in a simple way assures flexibility of predecision analysis since each reassignment to or from the list of critical resources corresponds to a simple redefinition of the decision problem.

With the above background **we can now define the task of IDS design or generating efficient development alternatives as a quest for concordance between available resources and technologies.**

The state of concordance is to be evaluated along well defined rules and measures for evaluation (and selection) of the efficiency of achieving goals (outputs) from resources supplied to industrial structure (inputs). Such rules and measures form a model of efficiency evaluation which is to be established in order to solve the quest for concordance to yield a feasible development alternatives. This problem area is dealt with in the paper ([9]).

Technological repertoire, critical resources, constraints and other factors describing the problem (or a particular industrial situation), are to be mapped into second model of a technological network.

Since it is intuitively obvious that such a process is to be performed through the generation and analysis of multitude of alternatives and their selection, then a mechanism is to be provided that enables to handle the situation. The appropriate models and means for handling the problem of quest for concordance may be organized into a system which is called simply DSS or Decision Support System.

The above philosophy stands behind development of MIDA system and the methodology. In the next step we shall present the assumptions used and a decomposition of the development problem which were applied for practical implementation of the above philosophy.

The effective approach taken in MIDA methodology in the practical implementation of the idea of the quest for concordance can be presented as follows. With respect to the decision level of the programming development, MIDA locates the IDS design on a level which could be named an intermediate economy level ([3], [9]). It goes between a macroeconomic level and microeconomic or corporate level. The first one proves to be too aggregated; a single technology cannot be considered in the analysis. Therefore a selection and assessment of appropriate technologies cannot be done at the macroeconomic level. On the other hand, a corporate or enterprise level also proves to be inadequate. This level is too narrow and particular to comprise complex technological and economic relationships which interact in the development process in the industry.

By identifying, defining and choosing the intermediate level, as the operational one for development programming, an important original feature has been decided in MIDA development. It comes together with the choice of an entity for setting the feasible scope of an industrial development problem which may be regarded as a basic object of the decision analysis. It is called PDA or Production Distribution Area. It responds to the necessity of a formal model of an industrial structure (IS) of the chemical industry. In fact, this model is formally described in [4] (see also appendix).

Due to a possibility of simple aggregation and disaggregation of the elements described in terms of the PDA model, also the PDA level can be split into several levels along so-called problem hierarchy ([3],[5] see Appendix). **This assures flexibility of the analysis** and well **corresponds to the industrial practice allowing at the same time to apply MIDA on different levels of aggregation (with data appropriate to the level considered).** It makes the concept and application of the PDA model very flexible and rather universal.

From the process of quest as considered so far on the PDA level, a goal structure can be selected representing the assumed state of IS at the end of the horizon covered by the analysis. However, to complete the task of programming, development of a feasible way of transition from the present or actual IS to the selected, final IS is to be optimally selected. The transition is to take into account the following factors:

- technological and market priorities,
- location possibilities,
- construction potential capabilities,
- availability of investment.

In short, to consider these factors, the investment necessary for the transition must be allocated both in space and time.

Therefore three levels emerge and provide a mentioned above decomposition ([7]) which was assumed in MIDA:

- selection of the final or goal IS,
- space allocation of investment,
- time allocation of investment (or investment scheduling).

The illustration of the MIDA methodology, following from decomposition applied in the IDS is given on the figure 1. Appropriate feedbacks between these levels provide through the space and time allocation a feasibility analysis of the goal structure originally selected.

The three level hierarchy and specifically, the space allocation and investment scheduling levels are discussed both theoretically and practically (through example) in ([9]).

It must be underlined at this point, that in fact **the decomposition of the IDS applied in MIDA approach corresponds to the managerial practice.** On the other hand, it can be conceptually opposed to more theoretical approaches based on dynamic programming which in general aims at global solution to be obtained from one model (see [2], [6]).

The general approach applied in MIDA follows from the common decision practice. First the goal "what" must be selected, then questions "where and how" should be answered. The decisive factor here is that the spatial allocation demands more detailed information related to sites and this must be confronted with spatially disaggregated values of critical resources as obtained from global solution. Site specific constraints must be also obeyed.

One more methodological disadvantage comes from globally formulated and solved problems — difficulty of interpretation — especially of cause-effect type. Too many factors are involved at once to enable that kind of analysis. It makes a real interaction with decision maker rather illusory.

5   Experience from applications

Major MIDA applications

For the scope assumed in this presentation of MIDA system and methodology some most important and representative applications were selected. The list of applications to be discussed

MIDA - Multiobjective Interactive Decision Aid

A methodology for integrated development programming

of the chemical industry

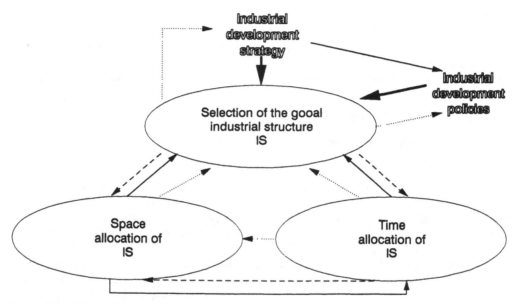

Industrial development strategy

- import substitution
- export promotion
- resource utilization
- heavy vs. consumer industry
- labor vs. capital intesive industry
- energy intensive vs. saving

Industrial development policies

- industrial tax
- rate of interest
- customs and duties
- individual vs. social income policy
- subsidies and special support funds

Fig. 1

is as follows:

1. **Polish Government Energy Program** — MIDA was used to elaborate a strategy for integration of energy and chemical sectors. MIDA study contributed to the fact that a new development, namely energochemical processing of coal, was brought to light and attained its place in the long-term policy.

2. **UNIDO projects**. JSRD competed successfully in offering its services to UNIDO and performed the following projects:

   - Master Plan for Development of the Chemical Industry in Iran,
   - Master Plan for Development of the Chemical Industry in Algeria,
   - Master Plan for Development of the Petrochemical Industry in Algeria.

   Within the framework of the above projects the services covered:

   - delivery and installation of equipment and adjacent software as well as delivery and installation of MIDA Decision Support System,
   - training of the counterpart personnel (using lectures, video tapes, top executive seminars and most of all *learning by doing* methodology),
   - elaboration of the development program in various alternatives,
   - industrial and system analysis consulting.

3. **Shanxi Case Study** — this application was done as a part of ACA project in IIASA for Shanxi Province in People's Republic of China and services performed were similar to those described for UNIDO, but the DSS software was developed as a spatially oriented version of the models incorporated in MIDA. The development program for coal based chemical industry was elaborated for the Shanxi province and technical expertise was also shared with the counterpart.

4. **SADCC Study** — *Study of the manufacture of industrial chemicals in the member states of SADCC* — this application was done under the contract with a consulting firm. The firm was contracted for a UNIDO project for SADCC countries. SADCC stands for Southern African Development Coordination Conference. Its members are 9 countries: Angola, Botswana, Lesotho, Malawi, Mozambique, Swaziland, Tanzania, Zambia, Zimbabwe. The consulting company contracted JSRD to perform application of MIDA system for the above study.

5. **Workshop on Integrated Development of the Chemical Industry**, organized by UNIDO for Brazil. This kind of workshop on IDP is based on the series of lectures and PDA case studies performed by participants using MIDA system. Learning by doing methodology is applied. Although this project has taken place after the UDINE School but was included in the paper due to the specific character of MIDA application; similar cases are to follow.

The above applications can be categorized to show range of problems and areas that can be tackled with a DSS and methodology such as MIDA as well as to provide a useful generalization of experience gained.

First category

A problem area related to the development of industrial sectors such as chemical and energy industries is selected. A research is to be carried out and forecasts provided with various technological and development alternatives. This is kind of predecision analysis which includes both research and application type of activities. The responsibility of JSRD as a contracted party covers all the work and study that may be considered as a kind of long range research programs with step by step results to be produced in form of progress and final reports. Results are used by various governmental agencies as well as other scientific centers.

This kind of application is exemplified by no 1 on the above list.

Here a DSS is used by the team performing the job mainly as a laboratory tool. No clearly defined decision maker is present in the process. In such case a variety of skills and experience, specifically presence of industrial experts in the team is especially decisive for good results to be obtained. In such cases by in parallel promoting a work devoted to the problem and a work done on developing methodology and DSS system proves to be fruitful and effective. Such in fact is organization of work assumed by JSRD.

Second category

A development program is to be elaborated for a foreign partner. Such were the applications that were contracted by JSRD with UNIDO. This covers wide span of services and responsibilities. The period assigned for the job is relatively very short : in the range of 1,5 — 3 month.

The DSS is to be delivered and installed together with computer equipment. Moreover, a user's team must be trained in a variety of skills including not only operation of DSS but first of all methodology of its application. These circumstances impose variety of demands which for the lack of space and the type of paper cannot be discussed in details but must be of deep concern. They can be briefly presented as follows.

The principle of operation of a DSS and methods applied should be as clear as possible and as simple as possible at the same time they must eliminate omitting or loss of any essential factors.

A great attention in DSS architecture, functioning and methodology must be paid to facilitate procedures which may help in validating both : simple source data and resulting development alternatives.

Users' involvement is a key factor, both to assure obtaining valuable and useful alternatives that would be accepted for implementation and to establish self-reliance of the users' team (including a decision maker). This can be done through very extensive educational effort and specifically by working out a "learning by doing" methodology. This must be backed also by very clear and well edited documentation supporting all activities as well as results of the project.

If one would like to compare the two above categories of application it could be formulated as follows.

First category provides more scientific and broad approach but is much less demanding in terms of software development, methodology and reliability of the system. On the other hand the second category provides extremely heavy duty testing of all elements taking part in the project.

This includes also all skills and abilities of people involved. It also provides important insights coming from different cultural and decision environments.

Moreover it provides also very useful cases which are an inspiration for the future developments in all aspects : theoretical, software and methodological.

Third category.

The system and methodology are to be adopted for different environment and are to be embedded in another system. Such is the case of the Shanxi case study a work done for ACA IIASA project contracted with Peoples Republic of China.

Apart from the previous remarks formulated for the case of UNIDO projects which remain valid, some additional observations can be formulated.

A DSS becomes a module of a larger system. All kinds of problems of interfacing with other types of software arise. The same concerns interfaces with other models.

At the same time in this particular case new elements specific for spatial allocation backed by scheduling of investment were also developed. In general this kind of applications help finding another way to generalization and standardization of architecture and functionality of the DSS not to mention new theoretical and methodological developments which usually also come in dealing with new, original problems.

Fourth category

This is a specific one when there is no direct contact with the field. The interface comes through third party. It provides a very useful kind of verification of system and methodology. It was the case of the fourth application listed above.

The experience gained so far from a single case reported here may be too limited to be generalized but due to difference in approaches and experience represented by the third party which is supposed to be professional in the field of programming development, a new light can be brought on the own approach which has to defend itself in such circumstances. In fact it also helps to test and improve system and methodology with procedures for validation of data and results.

Fifth category

Though exemplified here explicitly by one project in fact it is an important area of many applications in a number of countries including India. Moreover in each project a component of learning and doing was always present. The teaching environment proved to be a very challenging one and helped to improve methodology, especially in terms of conveying konwledge on rather specific, professional topics to participants representing a multidisciplinary group.

The above remarks summarize briefly experience in the domain of DSS as gained from major, categorized for that purpose applications of MIDA. Generality of categorization as well as of the relevant experience prove to be useful not only for a specific DSS such as MIDA.

6   Final remarks

MIDA or Multiobjective Interactive Decision Aid was used here to highlight various aspects and problems of decision support. Being on one hand a particular DSS aimed at application in a particular management area namely IDP — Integrated Development Programming, on the other hand proves to be a good platform for general observations that can be shared well beyond immediate field of its origin and application. This involves the problem of identification. It led to a counter-intuitive from the general theory point of view decomposition of the complex dynamic problem. Consequently it also led to the development also of the MIDA system which evolved in time with its core being PDA or Production Distribution Area model. In the system the data availability (to provide the user with detailed descriptive information) is of equal importantce with modeling capability (to provide him with normative information). It is assumed therefore that both categories of information are important in the process of learning and experimenting on the way to selection of a decision alternative. A great variety of options which can be accessed through the hierarchical menu of MIDA enables for a good interaction between user and problem performed in terms which come from normal practice of experts and decision makers. This is a feature which should be common in any DSS application. Lessons from various categories of applications show that DSS should be considered both as a laboratory tool for development of theory and methodology of decision support but also as a crude means in dealing with real decision problems. It simply shows that design, implementation and application is an open-ended process in which feedbacks from experience assure that this learning by doing process is both creative and rewarding in its scientific and practical effects.

Appendix

Basic model of industrial structure

Introduction

This appendix provides a reader with complete description and information on the model, formulas used as well as its linkage to industrial structure as identified in the IDP ([4]). The primary version of the model called PDA — Production Distribution Area was published in [1]. Currently presented version — stabilized formally few years ago as a result of its extensive use — constitutes a formal base for the MIDA.

Such a role of the PDA model in the whole activity stems from its open character and a way of its formulation. The openness means here that:

- a final shape of an operational PDA depends on a chemical branch modeled (data for the model),

- instead of a criterion or criteria, formulas named below the aggregates are proposed with possibility to combine them and to obtain complete optimization problem,

- there is a freedom in creation of constraints that may be put on the aggregates also,

- to formulate other new aggregated it is possible basing on the lower layer of the model that reflects static behavior of a network representing a strongly connected production structure of the chemical industry.

The model in its type is a quasi-linear programming one because some optimization scenarios are possible using pure linear formulas:

- a single objective case with a choice of criteria,

- a multi objective case,

- a linear-fractional case for a single objective.

Elements of the model

There are five types of elements of the model:

1. Installations - indexed by set I,

2. Processes - indexed by set P,

3. Media - indexed by set J,

4. Markets - indexed by set M.

5. Special resources - indexed by set S.

To reflect the possibility of running a number of different chemical processes on the same hardware the element called installation was introduced. Processes of the particular installation are dependent on a common capacity constraint.
The idea of installation splits I/O flows into two categories called:

media - flows which are interchanged between processes themselves and between processes and environment of the PDA, i.e. raw materials, products and by-products.

special resources - flows which are interchanged between installations and environment (common for all processes running on a particular installation), e.g.. investment, manpower.

To model an interface between the PDA and its environment, markets for media are introduced. The markets introduce characteristics of I/O media in terms of their sellability and availability. There may be up to four markets for a particular medium: domestic sale, domestic purchase, foreign sale (export) and foreign purchase (import). Because the category of special resources is not numerous they are treated separately.
Having in mind that media constitute a majority of flows, nobody ought to be surprised that it is a means for modeling not only particular chemicals, energy carriers but also pollutants in bulk (solid waste, liquid waste, emission) or in groups of toxic substances.

Formulas

Let us introduce variables of the model:

z_p, $p \in P$ - a level of production of process p,
y_j^m, $j \in J$, $m \in M$ - an amount of medium j bought or sold on market m.

An auxiliary equation describes amount of a given medium produced or consumed.

$$y_j = \sum_{p \in P_j^+} b_{jp}\, z_p - \sum_{p \in P_j^-} a_{jp}\, z_p \ , \quad j \in J \tag{1}$$

while the symbols above are defined as follows:
P_j^- - a subset of processes where medium j is consumed,
$a_{jp}\, z_p$ - quantity of medium j consumed (consumption coefficient),
P_j^+ - a subset of processes where medium j is produced,
$b_{jp}\, z_p$ - quantity of medium j produced (yield coefficient).

Balance equation for media

$$\sum_{m \in M_j^+} y_j^m - \sum_{m \in M_j^-} y_j^m = y_j \ , \quad j \in J \tag{2}$$

where:

M_j^+ - a subset of markets where medium j may be sold,

M_j^- - a subset of markets where medium j may be bought,

Balance of special resources

Formulas for a balance of the special resources have the same shape and only interpretation of coefficients is different.

$$Q^s = \sum_{i \in I} \sum_{p \in P^i} q_p^s \, z_p \ , \quad s \in S \tag{3}$$

where: P^i - the subset of processes running on the installation i.

For each special resource the coefficient q_p^s is estimated as a value related to the capacity of the process p .

The following special resources are defined in the PDA model:

- investment in a portion counted in the convertible currency, that must be paid abroad (the coefficient is a unit investment cost),

- investment in a portion counted in the local currency spent locally (unit investment cost for local expenses),

- labor — unskilled workers (a ratio men to capacity),

- supervision,

- labor — laboratory & control.

The applied above approach means that dependencies of investment and labor with respect to the capacity are linearized. To sustain reliability of the model at this point, a procedure for adjustment is assumed to keep an obtained production level reasonably close to the capacity.

Other special resources can be introduced for any special applications of the model.

Constraints on flows

Market constraints

$$\underline{y}_j^m \leq y_j^m \leq \bar{y}_j^m \ , \quad m \in M_j^+ \cup M_j^- \ , \quad j \in J \tag{4}$$

Special resources constraints

$$\underline{Q}^s \leq Q^s \leq \bar{Q}^s \ , \quad s \in S \tag{5}$$

Capacity constraints for processes

$$\sum_{p \in P^i} \frac{1}{\bar{z}_p} z_p \leq 1 , \quad i \in I \tag{6}$$

where: $\bar{z}_p$ - production capacity of the process p.

When a number of processes are to run on the same installation, the capacity is calculated under an assumption that the particular process occupies the whole installation.

Reconstruction constraints

$$\sum_{p \in P^o} \frac{1}{\bar{z}_p} z_p + \sum_{p \in P^n} \frac{1}{\bar{z}_p} z_p \leq 1 , \quad o, n \in I \tag{7}$$

Indexes o and n denotes an installation to be reconstructed and an installation after reconstruction, respectively. The constraint is defined for all couples of installation specified. There is a linearization of an exclusion condition that the old and new installations can not run simultaneously. Such a trick is successful when links of the installations in the network are similar.

Aggregates

Production value

$$\sum_{j \in J} \sum_{m \in M_j^+} c_j^m y_j^m \tag{8}$$

where: c_j^m - a price of medium j on the market m.

Cost of materials

$$\sum_{j \in J} \sum_{m \in M_j^-} c_j^m y_j^m \tag{9}$$

Another aggregate can be obtained as a difference of the production value and the cost of materials that is called the manufacturing value added.

Production cost

$$\sum_{i \in I} \sum_{p \in P^i} \alpha_p z_p + \sum_{j \in J} \sum_{m \in M_j^-} c_j^m y_j^m \tag{10}$$

where: α_p - a unit process cost coefficient for process p. The coefficient includes all cost factors modeled except the cost of materials. There are the cost of special resources and — depending on an assumed scheme of cost calculation — depreciation, taxes, assurance, overheads, etc.

The production cost can be used to get a formulas for profit and in turn for simple rate of return.

Energy equivalent of production

$$\sum_{j \in J} \sum_{m \in M_j^+} e_j^m \, y_j^m \tag{11}$$

where: e_j^m - unit energy equivalent of medium j.

For media that are not an energy carriers the lower heating value is used as the coefficient.

Energy equivalent of materials

$$\sum_{j \in J} \sum_{m \in M_j^-} e_j^m \, y_j^m \tag{12}$$

Using the two above aggregates derivative ones can be obtained: the energy lost in the system and the energy efficiency.

References

1. Borek, A., Dobrowolski G., Zebrowski M. (1978) GSOS — Growth Strategy Optimization System for the Chemical Industry. In Advances in Measurement and Control MECO'78, pp.1128-1131, Acta Press, Athens.

2. Dobrowolski, G., Rys T. (1981) A Dynamic Model for Development Programming of Production — Distribution Area. Theory and Applications. Materials of School of Economic Systems Simulation, Trzebiatowice (in Polish)

3. Dobrowolski, G., J. Kopytowski, T. Rys, M. Zebrowski (1985) MIDA - Multiobjective Interactive Decision Aid in the Development of the Chemical Industry. Theory, Software and Test Examples for Decision Support Systems, A. Lewandowski and A. Wierzbicki eds.,IIASA, Laxenburg, Austria.

4. Dobrowolski G., Żebrowski M., (1989) – *Basic Model of an Industrial Structure*, in Lecture Notes in Economics and Mathematical Systems No 331, Lewandowski, Wierzbicki Eds., Springer Verlag, Berlin 1989. pp. 287-294.

5. Gorecki, H., J. Kopytowski, T. Rys, M. Zebrowski (1984) Multiobjective Procedure for Project Formulation - Design of a Chemical Installation. In: M. Grauer, A.P. Wierzbicki (Eds.) Interactive Decision Analysis - Proc. of Int. Workshop on Interactive Decision Analysis and Interpretative Computer Intelligence Springer Verlag pp. 248-259.

6. Kendrick, D. A., A.J. Stoutjestijk (1978) The Planning of Industrial Investment Programs. Vol. 1. A Methodology. Johns Hopkins University Press, Baltimore, M.D., for World Bank Research Publications.

7. Kopytowski J., Żebrowski M. (1989) – *MIDA: Experience in Theory, Software and Application of DSS in the Chemical Industry*, in Lecture Notes in Economics and Mathematical Systems No 331, Lewandowski, Wierzbicki Eds., Springer Verlag, Berlin 1989 pp. 271-287.

8. Skocz M., Żebrowski M., Ziembla W. (1989), *Spatial Allocation and Investment Scheduling in the Development Programming*, in Lecture Notes in Economics

and Mathematical Systems No 331, Lewandowski, Wierzbicki Eds., Springer Verlag, Berlin 1989. pp. 322-339.

9. Żebrowski, M. (1987) Multiobjective Evaluation of Industrial Structures. MIDA Application to the case of the chemical industry. Theory, Software and Testing Examples for Decision Support Systems, A. Lewandowski and A. Wierzbicki eds., WP-87-26, IIASA, Laxenburg, Austria.

10. Żebrowski M., (1989) – *Multiobjective Evaluation of Industrial Structure*, in Lecture Notes in Economics and Mathematical Systems No 331, Lewandowski, Wierzbicki Eds., Springer Verlag, Berlin 1989. pp. 294-310.

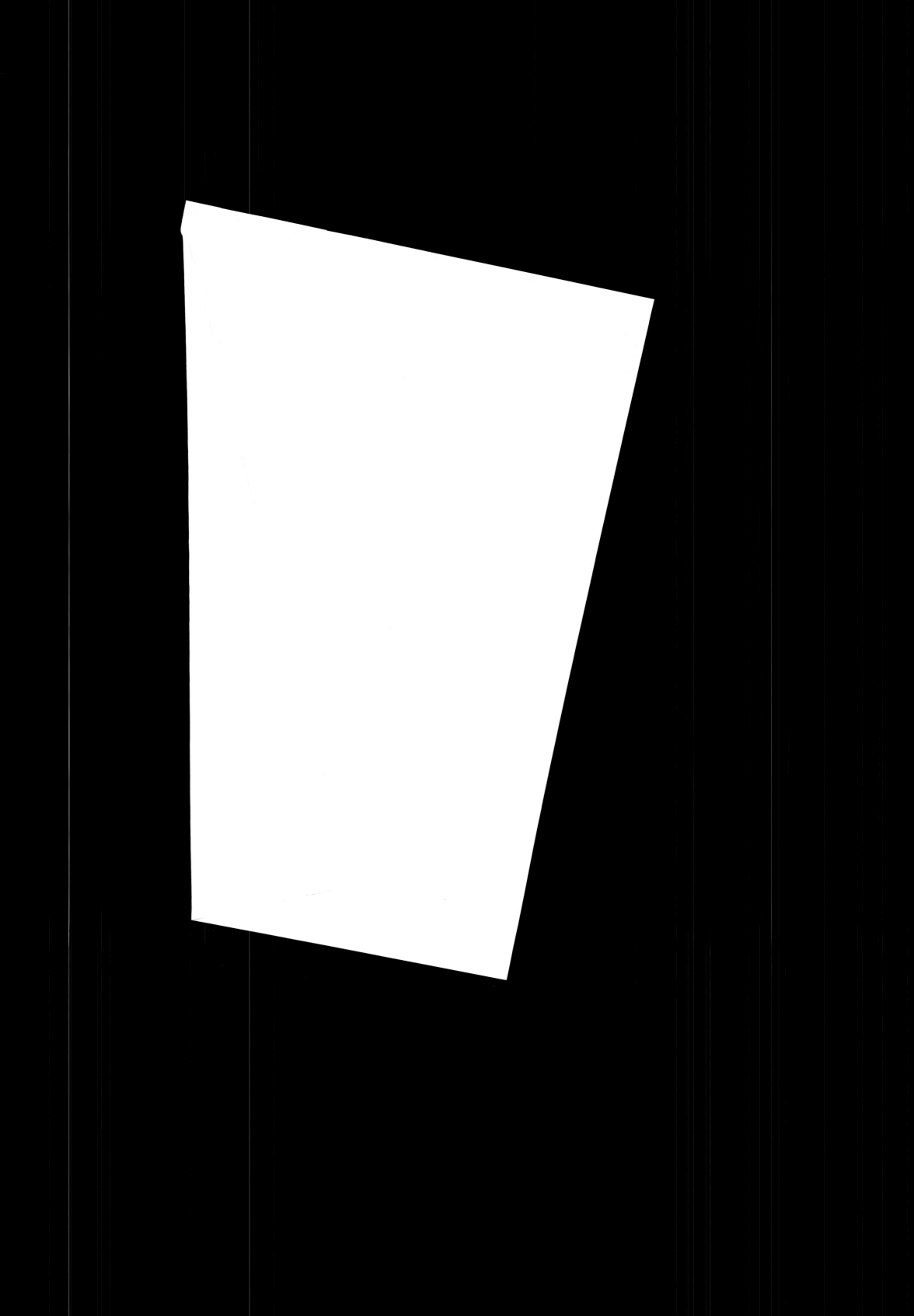

Printed in the United States
By Bookmasters